European Hair, Eye, and Skin Color: Solving the Puzzle

Peter Frost

European Hair, Eye, and Skin Color: Solving the Puzzle

Peter Frost

Academica Press
Washington~London

Library of Congress Cataloging-in-Publication Data

Names: Frost, Peter (author)
Title: European hair, eye, and skin color : solving the puzzle | Peter Frost
Description: Washington : Academica Press, 2022. | Includes references.
Identifiers: LCCN 2022943514 | ISBN 9781680538724 (hardcover) | 9781680538731 (e-book)

Contents

Figures

Tables

Preface

A rather glaring phenomenon has remained hitherto unexplained and unexplored, namely the strikingly greater heterogeneity of European populations (apart from recent, post-colonial migrations) in skin, eye and hair pigmentation than in other parts of the world such as Asia, sub-Saharan Africa, and indigenous America and Australia. Peter Frost presents the most impressive amount of data and hypotheses to account for this, largely in terms of sexual selection. In the final chapters, he explains how the original meaning of skin color became transformed and elaborated upon, as humans entered increasingly complex cultural environments marked by differences of class and ethnicity. Further on, he touches on gene-culture coevolution, and its role during the transition from the Mesolithic to the Neolithic in early Europe. A strikingly novel look at ourselves!

Pierre L. van den Berghe
University of Washington

Foreword

"*J'sais pas ... parce que je l'aime !*" The little boy had paused before answering, as if lost in thought. Another preschooler reached out with his forefinger, stroked the doll's bare legs, and then turned to me with a smile on his face.

It was the first year of my doctorate, and I wanted to study how young children perceive minor differences in skin color—the kind that would not seem ethnic or racial, such as those between men and women. For this I went to various daycare centers and showed each child a pair of dolls, one slightly darker than the other, and the child had to choose. An assistant then wrote down the choice, plus the child's age, sex, height, weight, adiposity, and skin reflectance under the upper arm.

Next came the analysis. Sex did not affect doll choice. Nor did height, weight, or skin reflectance. However, body mass index had a definite effect among the children less than three years old, as did triceps fat adjusted for body size. The darker doll was preferred by the ones with more body fat.

A professor in my department would later attack me for "seeking only those results I had expected to find." Well, no. The results came as a surprise. I had thought that doll choice might be associated with age, physical maturation, or awareness of one's own skin color. Or perhaps all three.

Though unexpected, the results fascinated me. I wanted to learn more. How could body fat be influencing this choice? What was going through the child's mind? I repeated the experiment with more children and began asking them afterwards why they chose one doll and not the other. But none of them knew.

The search for answers took me from physical to cultural anthropology … to a series of interviews with elderly men and women on Île aux Coudres, a rural community in eastern Québec. I expected some reticence, as an outsider with an accent and as an anthropologist with weird questions. Not to worry! They talked and talked at great length, and it was hard to bring each interview to an end. I soon understood the reason: they were lonely and welcomed any opportunity for human contact.

Yet sometimes they, too, had little to say. Not out of embarrassment. My assistant later explained: "Nowadays we're accustomed to asking questions about anything, and we think that's normal, but it wasn't that way in the past." My informants grew up at a time when talk covered a limited range of topics, with everything else being accepted as the way things were. So they had little to say about a topic they had never discussed before. They lacked the words. Or the ideas.

There was the same incomprehension among the professors in the anthropology department. At best, they did not know what to think. One of them wryly asked: "*Tu penses comprendre ces choses en regardant sous le bras d'un enfant?*" I nonetheless had to go where my curiosity led me. Outside the department a few people took interest in my work, notably Pierre van den Berghe at the University of Washington. My doctoral dissertation became the basis for several journal articles and finally a book published by the press of my university.

In time my thoughts turned to the question of European hair, eye, and skin color. How did these color traits come to be? Why are they so bright and diverse? Why do they all appear within the same geographic area? I wrote a paper arguing that relaxation of natural selection could not explain the unusually pale skin of Europeans. Later, I extended that enquiry to hair and eye color, eventually collaborating with Karel Kleisner and Jaroslav Flegr of Charles University in Prague.

To date, reactions have been mixed. Many people are intrigued by the idea that the spread of humans northward resulted in fewer men being available for mating and, thus, in women undergoing sexual selection. But why, they ask, was the selection stronger in Europe than in north Asia? Other people react negatively to the theory of sexual selection itself, calling it a catchall explanation for anything and everything. Finally, some argue that Middle Eastern farmers largely replaced European hunter-gatherers during Neolithic times. The European phenotype must have therefore appeared long after the ice age.

Debate will likely end with progress in two areas of research. One is ancient DNA. By retrieving DNA from human remains, we can see how people looked at different times of prehistory. We have already learned that the modern European phenotype—pale skin with a variety of hair and eye colors—already existed nine thousand years ago in the northeast of Europe. Other methods indicate that it came into existence no earlier than twenty thousand years ago. So this phenotype arose over a short time: less than ten thousand years, perhaps much less.

The second area of research is sex linkage, specifically the sexual differentiation of hair and eye colors during fetal life. In short, these colors are more evenly distributed among women than among men, apparently because of different degrees of exposure to estrogen in the womb. This hormone thus seems to favor fetal development of less common colors, like red hair, blond hair, and green eyes.

But I'm getting ahead of myself. Let me end my foreword here.

Summary

Europeans are strangely colored, particularly in the north and east. Hair is not only black but also brown, flaxen, golden, or red. Eyes are not only brown but also blue, gray, hazel, or green. Finally, skin is white, almost like that of an albino.

That phenotype is due to alleles at unrelated genes: *MC1R* for hair color, the *HERC2-OCA2* region for eye color, and *SLC45A2*, *SLC24A5*, and *TYRP1* for skin color. All of these alleles are relatively recent, having appeared among Europeans ten to twenty thousand years ago—long after the arrival of modern humans in Europe.

The European phenotype is also more sharply defined in women than in men. Women have a more even distribution of hair and eye colors: they have the less frequent ones more often and the more frequent ones less often. Women are especially overrepresented among redheads and blondes, while being three to five times underrepresented among people with black hair. Depigmentation of the skin has also gone farther in women than in men.

In each case, the sex difference is expressed differently. In the case of hair color, women less often have the darkest hue (black). In the case of eye color, they less often have the lightest hue (blue) and the darkest (brown). In the case of skin color, women are fairer than men because their skin has less melanin and less blood. These sex differences all share the same immediate cause: the degree of estrogenization during fetal development.

The ultimate cause seems to be sexual selection. If we consider the new colors that arose in Europe, we see that they are mostly on or near the face—the focus of visual attention. They are also brighter and "purer" than the original black hair and brown eyes; that is, they reflect more light and occupy thinner slices of the visible spectrum. Brightness and purity are characteristic of colors favored by sexual selection, as is novelty. A novel color improves mating success by attracting attention and interest, but that success is eventually its undoing. As each generation passes, it becomes more common … and less novel. The center of interest shifts to other colors, especially new ones that arise through mutation. Thus, over time, a growing number of color variants will accumulate within the population, in a dynamic equilibrium. This is how color polymorphisms come to be, and they are a frequent outcome of sexual selection.

Unlike the hair and the eyes, the skin did not develop a color polymorphism among Europeans, instead becoming unusually pale. The reason may be that sexual selection was guided by a pre-existing dimorphism. In all human populations, men are browner and ruddier than women, who by comparison are fairer. Fairer-skinned women were seen as more feminine in traditional cultures and were more sought after. Sexual selection, if sufficiently strong, would end up draining the gene pool of alleles for dark skin. This is likely what happened to early Europeans: their skin became pale, while their hair and eyes took on a diverse palette of new hues.

We know that Europeans initially had dark hair, dark eyes, and dark skin. Their current appearance emerged much later—during the last ice age ten to twenty thousand years ago, when their diet was largely meat from reindeer and other migratory game. Long-distance hunting increased the death rate among men and decreased the polygyny rate—only the ablest hunters could provide for more than one woman and her children because women had almost no food autonomy. The result: a surplus of women on the mate market; intense rivalry among them for male attention; and strong selection for eye-catching female features. Such features became more frequent with succeeding generations, eventually forming what is now seen as the European phenotype.

Introduction

Europeans are strangely colored, particularly in the north and east. Their hair is not only black but also brown, flaxen, golden, or red. Their eyes are not only brown but also blue, gray, hazel, or green. Finally, their skin is almost albino white.

Are these hair and eye colors a side effect of white skin? No, the genes are different in each case. Europeans owe their pale skin to new alleles at *SLC45A2*, *SLC24A5*, and *TYRP1*,[1] their multicolored hair mostly to new alleles at *MC1R*,[2] and their multicolored eyes to new alleles in the *HERC2-OCA2* region and elsewhere.[3]

Some of those alleles have general effects on pigmentation, but others do not. Hair and eye color can vary independently to a high degree.[4] This point was made by the American anthropologist Alice Brues:

> If we were to take all the human beings in the world who have dark brown eyes and black or dark brown hair, we would not only have the vast majority of the human species, but would have a group which shows virtually the complete range of human skin color, from black to almost completely depigmented.[5]

---

[1] Beleza et al. 2013; Canfield et al. 2014; Soejima et al. 2005; Voight et al. 2006. See also Table 1.

[2] Box et al. 1997; Flanagan et al. 2000; Makova and Norton 2005; Rana et al. 1999. See also Table 1. According to Branicki et al. (2011) *MC1R* has a predictive power of 86-88% for red hair and 63-68% for other hair colors. According to Morgan et al. (2018), *MC1R* explains 73% of SNP heritability for red hair. SNPs at *ASIP* have also been associated with red hair (Morgan et al. 2018). An SNP at *KITLG* is associated with blond hair (Ainger et al. 2017).

[3] Duffy et al. 2007; Eiberg et al. 2008; Kayser et al. 2008; Liu et al. 2010; Sturm and Frudakis 2004; Sturm et al. 2008. See also Table 1. An SNP at *HERC2* has a predictive power of 88% for blue eyes and 90% for brown eyes. It explains on average 46% of variance in eye color (hue and saturation) (Walsh et al. 2013).

[4] Guenther et al. 2014, p. 751.

[5] Brues 1975.

The same point is made by the authors of a recent genome-wide association study, which showed weak linkages between hair, eye, and skin color:

> Although DNA variants within the *MC1R* gene are strongly associated with light skin and red hair color, no detectable association with eye color was found in our large GWAS, in line with previous albeit smaller-sized GWASs of more limited statistical power. Similarly, other DNA variants strongly associated with skin and hair color within genes, such as *SILV*, *ASIP*, and *POMC*, showed no statistically significant effect on eye color in this study, nor in previous studies. Moreover, we also identified 34 genetic loci that were significantly associated with eye color, but for which there is no report of significant association with hair and/or skin color.[6]

Some alleles for hair or eye color can lighten skin color, particularly the ones for red hair or blue eyes, and they may indeed be a side effect of selection for lighter skin.[7] But why would such selection increase the total number of alleles for hair and eye color? And why have neither red hair nor blue eyes reached fixation in any human population, even those with milk-white complexions?

Two decades ago, researchers began identifying the alleles at *MC1R*, the main gene for hair color. They found large geographic differences: eleven nonsynonymous alleles among Europeans versus five among Asians and one among Africans.[8] These numbers actually understate the diversity of European hair color because the Asian alleles are almost synonymous (i.e., indistinguishable in their visual effect). Among Europeans, there are twice as many nonsynonymous *MC1R* alleles as synonymous ones, a ratio that is the reverse of what we see at most genes.[9] Something has diversified not only the nucleotide sequences for hair color but also, and even more so, actual hair color.

Two research teams looked into the question. The first team concluded that hair color diversified among Europeans through selection

[6] Simcoe et al. 2021.
[7] Duffy et al. 2004; Sturm and Frudakis 2004.
[8] Harding et al. 2000; Rana et al. 1999.
[9] Rana et al. 1999. Synonymous alleles produce the same phenotype.

for a wide range of colors.[10] In other words, the diversity itself was somehow advantageous. The second team disagreed, concluding that the cause was relaxation of selection. They argued that selection for dark skin grew weaker as modern humans spread out of Africa and into northern latitudes with weaker ultraviolet light and less risk of sunburn. The gene pool was no longer purged of defective alleles for pigmentation and could now accumulate alleles for new hair colors.[11]

The second scenario has been modeled by two geneticists, Alan Templeton and Rosalind Harding. They estimate that close to a million years would be needed to accumulate the hair color diversity we now see in Europeans, including approximately 80,000 years for today's prevalence of red hair alone.[12] Because modern humans entered Europe only 45,000 years ago, some science writers concluded that Europeans inherited their diverse hair colors from the Neanderthals, particularly red hair.[13]

The debate has taken a new turn with the discovery that the various hair colors are due to far more than a dozen alleles. Over two hundred have been found, for colors ranging from blond to black.[14] If these variant alleles had gradually accumulated because natural selection was no longer purging them, the process of accumulation would have taken tens of millions of years.[15]

---

[10] Rana et al. 1999.

[11] Harding et al. 2000. This scenario rests on the questionable assumption that hair color is strongly linked to skin color.

[12] Harding et al. 2000; Templeton 2002.

[13] Both Alan Templeton and Rosalind Harding seem to make this argument in their journal articles. Harding, in particular, has been quoted to that effect in the press. However, she disputes the press account, saying she was misquoted (Roach 2005, pp. 125-126).

[14] Morgan et al. 2018. There are probably many more than two hundred. Hysi et al. (2018), using almost 300,000 participants of European ancestry, identified 124 loci associated with hair color. Many of those loci would have several alleles.

[15] In a study of two SNPs at *MC1R*, Huang et al. (2021) concluded that "genetic drift cannot explain the genetic divergence of *MC1R* between Africans and Eurasians. Our statistical results show that the divergences of rs2228479 and rs885479 between Eurasians and Africans are highly significant departure from neutral evolution, suggesting that directional selection is the more likely explanation."

That span of time is too long even if it began with the earliest Neanderthals, who lived at most three quarters of a million years ago.[16] That is hardly enough time for a dozen new alleles to accumulate through relaxation of selection, let alone more than two hundred. Furthermore, analysis of five red-hair alleles at *MC1R* has shown that only one of them could be a Neanderthal introgression, the others being apparently of modern human origin.[17] Even if we assume a Neanderthal origin for all of the alleles for hair color, some kind of positive selection—and not just absence of selection—would have been needed to boost their frequencies to the levels they now have in the European gene pool, which is only 1 to 4% Neanderthal.

Meanwhile, eye color also diversified among Europeans (see Table 1).[18] This polymorphism, like the one for hair color, involves many alleles at many gene loci. To date, 124 have been identified, accounting for 53% of the diversity in European eye color. This diversity has multiple physical causes, such as differences in Tyndall scattering and melanin metabolism.[19] The ultimate cause must have been a process of selection that acted directly on the phenotype—something was favoring the reproduction of those individuals who had new eye colors. The selection seems to have been very strong: the main determinant of eye color, the *OCA2-HERC2* region, has the third-longest haplotype span of diminished heterozygosity that may be found in the genome of Europeans.[20]

In sum, two polymorphisms with different genetic and physiological causes, one for hair color and the other for eye color, arose within the same narrow limits of time and space. That was no accident. Some kind of selection was responsible. Early Europeans gained some kind of advantage from having a wide range of hair and eye colors.

---

[16] Rogers et al. 2017.
[17] Ding et al. 2014.
[18] Liu et al. 2010. Contrary to widespread belief, eye color does not follow a simple two-allele model of inheritance where brown eyes are dominant and blue eyes recessive. A single copy of the blue-eye allele usually produces an intermediate hue, like green or hazel, and even two copies do not always produce blue eyes (Branicki et al. 2009; Eiberg et al. 2008; Martinez-Cadenas et al. 2013).
[19] Simcoe et al. 2021.
[20] Voight et al. 2006, see also: Donnelly et al. 2012; Lao et al. 2007; Suarez et al. 2021.

Table 1: Evolution of European color traits

Gene	Estimated time of origin of European alleles	New phenotype
SLC45A2	19,000 – 11,000 BP (Beleza et al. 2013) ≈ 11,000 BP (Soejima et al. 2005)	Lighter skin
SLC24A5	19,200 – 7,600 BP (Canfield et al. 2014) 19,000 – 11,000 BP (Beleza et al. 2013) ≈ 12,000 – 3,000 BP (Norton and Hammer 2007)	Lighter skin
TYRP1	19,000 – 11,000 BP (Beleza et al. 2013)	Lighter skin
HERC2-OCA2	≈ 10,000 – 6,000 BP (Eiberg et al. 2008) before 9,500 BP (Günther et al. 2018) before 8,000 BP (Lazaridis et al. 2014) before 7,500 BP (Mittnik et al. 2018) before 7,000 BP (Olalde et al. 2014)	New eye colors
MC1R	no dating yet	New hair colors

Was it natural selection?

Perhaps natural selection created these polymorphisms of hair and eye color. We return here to the previous scenario but in another form. As modern humans moved into higher latitudes with weaker ultraviolet light, their skin may no longer have produced enough vitamin D. Natural selection would have then favored paler individuals whose skin could let in more UV for vitamin D synthesis. The new hair and eye colors were therefore a side effect.

Again, this scenario overstates the genetic linkages between hair, eye, and skin color. One may also object that the high ratio of non-synonymous to synonymous alleles is hard to reconcile with a simple side effect. But there is a more important objection. We know that Europeans became fair-skinned no earlier than 20,000 years ago—at least 25,000 years after modern humans arrived in Europe. This finding has been confirmed by three research teams.[21] As a *Science* correspondent concluded: "The implication is that our European ancestors were brown-

[21] Beleza et al. 2013; Canfield et al. 2014; Norton and Hammer 2007; see Table 1.

skinned for tens of thousands of years."[22] If natural selection made Europeans fair-skinned to alleviate a vitamin D deficiency, why did it take so long to act?

We can circumvent this problem by assuming that the deficiency began later, when farming replaced hunting, fishing, and gathering. This is the hypothesis of anthropologist Joseph Henrich: with the arrival of farming some six thousand years ago, people around the Baltic began to consume less fish and, hence, less vitamin D. So natural selection now favored individuals with lighter skin. Again, the new hair and eye colors are considered to be simply a side effect.[23]

That hypothesis has actually been tested. Two research teams have analyzed DNA from the remains of humans who lived in Scandinavia and the Baltic region during the Mesolithic, that is, DNA from the remains of people who had not yet begun to farm. Analysis showed that they already had white skin, as well as a variety of hair and eye colors.[24] The new phenotype therefore preceded the arrival of farming, being initially confined to the hunter-gatherers of northeast Europe.

Henrich's hypothesis is also very time-constrained. The new alleles—more than two hundred for hair color alone—would have had to appear and rise to their current population frequencies between the arrival of farming and the first ancient paintings that showed Europeans with pale skin and various hair and eye colors. That span of time could not be more than three thousand years. Such rapid change requires extremely intense selection. Is all of that compatible with Henrich's model? Or any model? Moreover, it's not as if people around the Baltic stopped eating fish altogether. That food source remained important for them well into the time of recorded history.

The scientific literature offers us another scenario of natural selection. Perhaps lighter-colored eyes were favored as a means to improve visual acuity in dim light, such as exists in the misty maritime climate of northwest Europe.[25] Hair color therefore became lighter as a

---

[22] Gibbons 2007.
[23] Henrich 2017, pp. 84-85.
[24] Mittnik et al. 2018; Günther et al. 2018.
[25] Short 1975.

side effect. Again, the scenario is at odds with the weak genetic linkage between eye and hair color. Furthermore, eyes are light-colored over large areas of Europe where the climate is dry and continental. Finally, it is unclear why this scenario would favor a variety of hues and not simply whichever one is best in dim light.

Was it sexual selection?

Perhaps sexual selection created the polymorphisms of hair and eye color. There is a *prima facie* reason for thinking so. Usually, a colorful appearance is not an adaptation to the natural environment, which favors a drab and unobtrusive look to avoid detection by predators.[26] Usually, a colorful appearance is favored by the social environment, as a means to improve your chances of getting noticed.

"Visual merchandising" is most important in saturated markets that offer too many interesting choices among products of equal quality.[27] The same principle applies to the mate market. If a sex difference in the death rate or the polygamy rate results in too many of one sex and too few of the other, the sex in short supply will get to pick and choose. They will tend to choose mates with eye-catching characteristics, and those characteristics will be passed on to the next generation. Thus, over many generations there will be an evolution toward flashier and more conspicuous features.[28] That is sexual selection.

Sexual selection favors brightness and novelty. A bright object stays in memory longer, and a novel one holds attention longer.[29] Brightness and novelty can also boost each other's attractive power, as in the case of an object that is bright within a thin slice of the visible spectrum. Because narrow-spectrum colors are rare in nature, Konrad Lorenz argued that they "release" strong feelings of interest:

---

[26] Kirkpatrick 1987.

[27] Lea-Greenwood 1998; Oakley 1990.

[28] Darwin 1936[1888], pp. 567-924; Emlen and Oring 1977.

[29] Brockmole and Boot 2009; Crothers and Cummings 2013; Hogben et al. 1976; Lee et al. 2011; Maan and Cummings 2009; Ossip-Drahos et al. 2016. "It would even appear that mere novelty, or slight changes for the sake of change, have sometimes acted on female birds as a charm, like changes of fashion with us" (Darwin 1936[1888], p. 813).

> Throughout the universe it is order that is improbable while disorder, chaos, entropy or whatever one chooses to call it, is what we should expect from the laws of probability. I do not think it too far-fetched an explanation to suppose that the striking orderliness and regularity which so strongly appeals to our sense of beauty in the coloring, the notes and the display movements of so many animals, especially of birds, has its source in their nature as releasers and in the general tendency of all releasers to develop in the direction of the more improbable. This would also explain the astonishing rhythm that we meet in very nearly every releasing action. The more or less pure spectral colors which so often appear in color patterns functioning as releasers, very probably also find their explanation in this way, since the reflexion of one wavelength among the wave-mixture of white light is in itself rather improbable, so improbable indeed that color alone may, in some cases, function as a releaser.[30]

It has been counter-argued that narrow-spectrum colors get noticed because they are mentally associated with edible fruits.[31] But that argument confuses cause and effect: edible fruits became vividly colored precisely to get noticed, as did flowers and venomous snakes.

If we return to European hair and eye colors, we see four indications of sexual selection:

Brighter hues

The new hues are brighter. Black hair and brown eyes have given way to a palette of bright colors, despite the possibility of new dark colors. Hair is carrot red, not beet red. Eyes are light blue, not navy blue. Brightness increases visual impact by causing the observer to watch the image longer and keep it in memory longer.

Purer hues

The new hues are "purer." Broad-spectrum colors have given way to narrow-spectrum colors, like blue, green, red, or yellow. In each case, brightness is concentrated within a thin slice of the visible spectrum—a few wavelengths. Pure colors do not happen by accident. They are unusual

[30] Lorenz 1939, p. 249.
[31] Rodd et al. 2002.

in the natural world and almost always serve to attract attention, either as a warning coloration or as a means to attract a mate.

More hues altogether

There are more hues altogether. A color polymorphism has arisen, most likely through attraction to novelty. A color gets noticed not only by being bright and pure but also by being rare. Over generations of mate choice, this novelty effect will exhaust itself for any one color, with the result that more and more variants will accumulate in the population via a recurring cycle:

1. A new color appears by mutation.
2. Its novelty value enhances the mating success of those who have it.
3. Over successive generations, it becomes more prevalent within the population, and less novel. Its bearers no longer enjoy the same mating success.
4. Its prevalence stabilizes, and a new equilibrium is formed among the various colors.
5. Another color eventually appears by mutation, and the cycle repeats. Over time, the population will collect a growing number of color variants.

A more balanced variety of hues

Finally, women have a more even distribution of these variants, having the less frequent colors more often and the more frequent ones less often. Red hair is especially overrepresented among women, followed by blond hair and light brown hair, whereas black hair is three to five times underrepresented.[32] A twin study has confirmed that hair color is more varied among women, who are especially prone to having red hair.[33] The different eye colors are likewise more evenly distributed among women.[34]

These sex differences are not simply a matter of one sex being generally less pigmented than the other. Yes, this is the case with hair

[32] Hysi et al. 2018.
[33] Shekar et al. 2008.
[34] Frost et al. 2017. See also Table 2.

color: women are underrepresented among individuals with black hair and overrepresented among those with red, blond, or light brown hair. [35] But this is not the case with eye color: women are underrepresented among individuals with blue eyes or brown eyes and overrepresented among those with green eyes. [36] The sex difference has different immediate causes in each case.

Nonetheless, these two color polymorphisms seem to share a common developmental cause, not only with each other but also with other sexually dimorphic features that are better defined in women than in men. A Czech study found a more feminine face shape in blue-eyed men than in brown-eyed men, as if a single factor had feminized both face shape and eye color, most likely during fetal development.[37] A hormonal cause is consistent with the sexual differentiation of hair color at puberty, when girls become lighter-haired than boys.[38] During adulthood, blond hair darkens with age more slowly in women than in men.[39]

Sexual selection and rare-color advantage

If we look at other species, we see that sexual selection has created many polymorphisms of bright, pure hues, notably in guppies, lizards, and insects.[40] In addition, a number of bird species have color polymorphisms

---

[35] Shekar et al. 2008.
[36] Frost et al. 2017; Martinez-Cadenas et al. 2013.
[37] Kleisner et al. 2010; Kleisner et al. 2013.
[38] Steggerda 1941.
[39] Olivier 1960, p. 74.
[40] Anderson 1969; Brooks 2002; Eakley and Houde 2004; Endler 1980; Farr 1980; Grant et al. 1974; Hampton et al. 2009; Healey et al. 2008; Hughes et al. 1999; Hughes et al. 2005; Kokko et al. 2007; Muggleton 1979; Olendorf et al. 2006; Simchuk 2001; Sinnock 1970; Tong et al. 2009; Wellenreuther et al. 2014; White 2017. Color polymorphisms would be more widespread in the animal kingdom if animals did not have to contend with (1) predation pressure, this being a constraint on color traits in general, and (2) the presence of related species within the same geographic range, apparently because too much intraspecific variability interferes with species recognition and leads to hybridization (Endler 1980). Neither constraint applied to early Europeans. Outside Africa, humans had only two potential predators: wolves and bears, the latter being uncommon and the former only an occasional threat to recent hunter-gatherers (Hoffecker 2002, pp. 238, 240). Hybridization was just as nonproblematic. The only other *Homo* population in Europe, the Neanderthals, had gone extinct ~40,000 years ago (Higham et al. 2014).

whose evolutionary origin is unclear.[41]

Nonetheless, evolutionary biologists are uneasy with the idea that a color can be attractive simply by being bright, pure, or novel. What is to be gained? Surely, natural selection would eliminate such a useless response. Yet that is easier said than done. Think of the trouble we have ignoring advertising techniques: the catchy jingles, the flashy look and feel of glossy promo ... Such techniques grab our interest by exploiting half-conscious needs; in this case, the need to monitor our surroundings for anything out of the ordinary. We cannot disable such mechanisms without exposing ourselves to possibly serious risks.

Rare-color advantage has been reported in humans, particularly as an advantage for sexual attraction. An American research team showed pictures of attractive women to male participants and asked them to choose the one they most wanted to marry. There were three series of pictures: the first had equal numbers of brunettes and blondes; the second had one brunette for every five blondes; and the third had one brunette for every eleven blondes. The result? The scarcer the brunettes were in a series, the more attractive they seemed—each brunette had a better chance of being chosen.[42] Other studies have reported similar findings. *Maxim* cover girls are more often light blonde or dark brown than the more usual dark blond or light brown of real life.[43] Viennese women tend to change their hair to a less common color.[44] In Brazil and the United Kingdom, modeling agencies prefer women with less frequent eye colors.[45] On American TV, women are four and a half times likelier than men to have red or auburn hair and five times likelier to have blonde hair.[46] On Turkish TV, women more often have red or blonde hair.[47] Rare-color advantage has also been observed in sales of home interior paints. Over the past half-century,

---

[41] Lank 2002.

[42] Thelen 1983. See also Pincott 2008, pp. 82-85.

[43] Anon 2008b. See also Figure 3.

[44] Schweder 1994.

[45] Forti and Young 2016.

[46] Davis 1990. Rare-color advantage may explain why blonde *Playboy* playmates have become more frequent since the mid-1960s, increasing from a third to over half of the total (Anon 2008a). The same period has seen blondes decline as a percentage of the American population.

[47] Ikizler 2007, p. 39.

preference for one paint color would rise until satiated and then give way to preference for another.[48] However, no such advantage was observed in a replication of the original American study. It should be noted that the replication differed from the original in two ways: participants were recruited online and made their choices privately on their home computers, with no control over whatever female images they had viewed or might still be viewing within sight of the computer screen.[49]

In sum, rare-color advantage, combined with strong sexual selection, may have diversified European hair and eye color. This scenario would explain two seemingly contradictory findings: although blue eyes show a strong signal of selection,[50] a German study showed no overall preference for them.[51] Initially, when blue eyes were rare, they must have been strongly preferred; today, they no longer are, and the various eye colors are in equilibrium. As the authors of the study note: "Perhaps the frequency of eye colors plays a role. In most countries, blue eyes are less prevalent than other eye colors and may have the image of something special and more valuable. If this assumption is true, brown eyes should be preferred in countries where the majority of the population has blue eyes."[52]

---

[48] Stansfield and Whitfield 2005.

[49] Janif et al. 2015. Rating of facial images is more problematic online than in a supervised laboratory. An online context provides no control over viewing conditions—not only image size and quality but also the presence of similar images nearby or even on the same screen. A man will often have pictures of women on his computer or iPhone. If a male rater has been viewing such images, his mental processing of the target image will be less focused.

[50] Donnelly et al. 2012.

[51] Gründl et al. 2012.

[52] Gründl et al. 2012.

Table 2. Frequencies of eye colors in men and women.[53] Less common eye colors are overrepresented in women.

	Blue	Brown	Green	Black	Gray	Amber	Hazel	Yellow
Men	33.40%	31.53%	18.52%	0.51%	12.16%	0.90%	2.94%	0.00%
Women	28.00%	26.76%	25.32%	0.56%	11.37%	1.44%	6.44%	0.08%
p	0.000	0.000	0.000	0.818	0.347	0.059	0.000	-

What about skin color?

Unlike the hair and the eyes, the skin did not develop a color polymorphism among Europeans, instead becoming unusually pale. The reason may be that sexual selection was guided by a pre-existing dimorphism: in all populations, men are browner and ruddier than women, who look fair by comparison.[54] Lighter-skinned women are thus seen as more feminine. When anthropologist Pierre van den Berghe surveyed the *Human Relations Area Files*, the main database of world cultures, he found a recurring preference in traditional cultures for women with skin lighter than average (see Table 3). The feminine ideal was therefore "white" in most of Eurasia, "golden" in Southeast Asia, and "yellow" or "red" in sub-Saharan Africa.[55]

If, within a population, sexual selection of lighter-skinned women is sufficiently strong, it will progressively drain the gene pool of alleles for dark skin. This is likely what happened to early Europeans, who became almost as pale as albinos.

As mentioned earlier, that change happened over a relatively short time and within a relatively small area. Europeans were initially brown and became fair no earlier than 20,000 years ago, and then only in the northeast of their continent.[56] To the west, skin remained brown almost into historic times.[57] It remained likewise darker to the east, even though

[53] Frost, Kleisner, and Flegr 2017.

[54] Edwards and Duntley 1939; Edwards and Duntley 1949; Edwards et al. 1941; Manning, Bundred, and Mather 2004; van den Berghe and Frost 1986.

[55] van den Berghe and Frost 1986; Frost 1988.

[56] Beleza et al. 2013; Canfield et al. 2014; Norton and Hammer 2007; see Table 1; Günther et al. 2018; Mittnik et al. 2018.

[57] Brace et al. 2019; Lazaridis et al. 2014; Olalde et al. 2014.

ground-level UV is equally low across Eurasia and North America above 47°N.[58] A recent genomic study has shown that selection for fair skin was about four times stronger among ancestral Europeans than it was among ancestral North Asians or among the earlier shared ancestors of both groups.[59]

Table 3. Preference for fair skin in women and men. Number of societies by geographic area[60]

Area	Women only	Both sexes	Men only	Doubtful or negative cases
Sub-Saharan Africa	9	2	1	0
Europe and Soviet Union	1	1	0	0
Asia	5	4	0	1
Middle East and Muslim Africa	2	0	0	0
North America	3	1	1	0
South America and Caribbean	5	2	1	1
Insular Pacific	5	4	0	2
Total	30	14	3	4

But why are women lighter-skinned than men? Since this sexual dimorphism exists in all humans, its origins likely go back to a time when our ancestors were not yet human. We must therefore turn to the literature of primatology and ethology.

A woman's body has a lighter complexion probably for the same reason it has other "cute" childlike features: a babyish face, a higher pitch of voice, and smoother, more pliable skin. Such features seem to promote pair-bonding by making men less aggressive and more caring.[61] Infants have lighter skin not only in humans but also in other primates, which

[58] Jablonski and Chaplin 2000, pp. 71, 76.
[59] Huang et al. 2021.
[60] Source: Human Relations Area Files (HRAF), Category 832 Sexual Stimulation, Ideals of Erotic Beauty and Sexual Attraction, see van den Berghe and Frost 1986.
[61] Frost 2010, pp. 134-135; Lorenz 1971, pp. 154-164; Wickler 1973, pp. 255-265.

apparently use this identifying mark to find wayward offspring. As the infant darkens with age, it no longer arouses the same interest from adults, including its mother.[62]

At some point in human evolution, women acquired a lighter coloration, along with other visual, auditory, and tactile aspects of the infant. That visual cue became a means to tell men and women apart.[63] In fact, when shown a photo of a face, subjects can use its hue and luminosity to identify the person's gender, even if the image is blurred and provides no other useful information.[64] The mental process seems unconscious.

In sum, lighter skin became mentally associated first with infants and later with women. Both mental associations were inevitable in earlier social environments, since age and sex differences in complexion accounted for most of the observable variation in skin color within any one society. In time, there arose a social norm: women were expected to have lighter skin, and they tried to make it even more so through various practices, like avoiding the sun, wearing broad-brimmed hats, applying facial powders, and so on.

In the post-Columbian world, the sex difference in complexion has become overshadowed by much larger differences of race and ethnicity. Furthermore, many women have turned away from fair skin to embrace the tanned look, thus giving their body surface a new erotic meaning.[65] This fashion, which began in the post-traditional Western world of the early twentieth century, has had trouble penetrating other societies, even highly Westernized ones like Japan.[66]

---

[62] Alley 1980; Alley 2014; Blaffer-Hrdy 2000, pp. 446-448; Booth 1962; Jay 1962.
[63] Bruce and Langton 1994; Hill, Bruce, and Akamatsu 1995; Russell and Sinha 2007; Russell et al. 2006; Tarr et al. 2001; Tarr, Rossion, and Doerschner 2002.
[64] Tarr et al. 2001.
[65] Segrave 2005.
[66] Ashikari 2005, pp. 85-86.

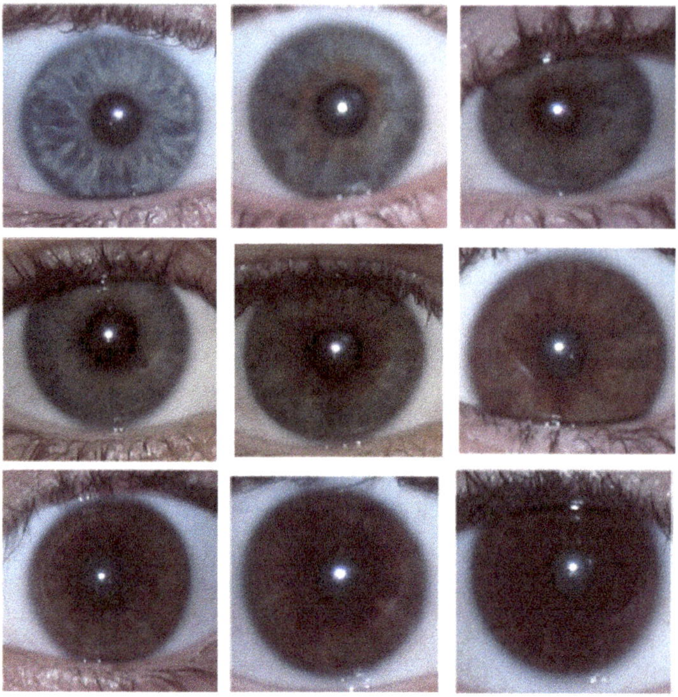

Figure 1: Palette of eye colors (Credit: Prof. Richard A. Sturm, Dermatology Research Centre, University of Queensland)

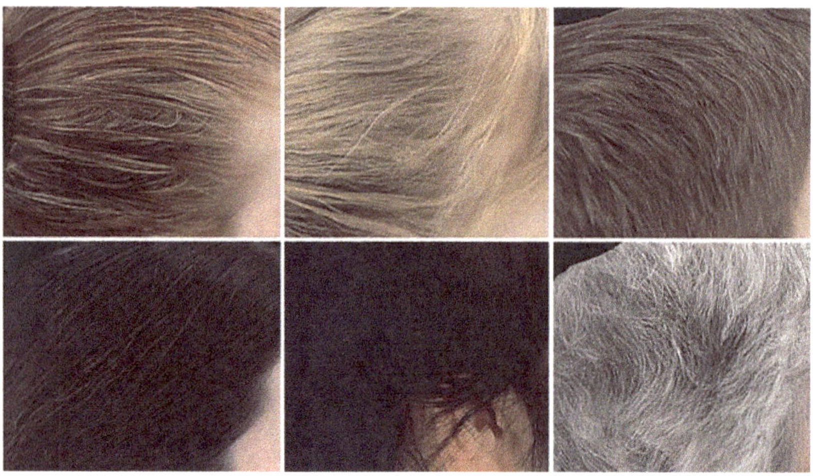

Figure 2: Palette of hair colors (Credit: Prof. Richard A. Sturm, Dermatology Research Centre, University of Queensland; Ainger et al. 2017)

Earlier thinking

Has anyone else attributed the unusual color traits of Europeans to sexual selection? A few have. One is geneticist Luigi Luca Cavalli-Sforza:

> [...] both hair and eyes are lightest in the south Baltic and their darkness increases fairly regularly and almost concentrically around this region. One may wonder whether this center was the place of origin of mutation(s) to light pigmentation and whether there may have been a cause for selection (natural or sexual) of light pigmentation in this region.[67]

Another is anthropologist Herbert John Fleure:

> In a few places in Sweden, Britain, and France people have been noticed who show characteristics of the skull and face that remind one of late-Paleolithic man: these people are usually darker, in hair and eyes, than their neighbors; sometimes they even have swarthy skins. Although this fact may not have great weight in argument, it does hint that there has been depigmentation in this region. The many stories of golden hair and blue eyes suggest that sexual selection may have helped the change.[68]

There is also science writer Jared Diamond:

> But there's an overwhelming importance to skin, eye, and hair color that is obvious to all of us—sexual selection. Before we can reach a condition of intimacy permitting us to assess the beauty of a prospective sex partner's hidden physical attractions, we first have to pass muster for skin, eyes, and hair.[69]

Although Charles Darwin was the one who came up with the theory of sexual selection, he never used it to explain European hair and eye colors, at least not explicitly. He did wonder whether that theory could explain how women in general became lighter-skinned. On the first page of "Sexual Selection in Relation to Man" he wrote:

> In certain races the women are said to differ slightly in tint from the men. For instance, Schweinfurth, in speaking of a negress

[67] Cavalli-Sforza, Menozzi and Piazza 1994, p. 266.
[68] Fleure 1945, p. 591.
[69] Diamond 1994.

> belonging to the Monbuttoos, who inhabit the interior of Africa a few degrees north of the Equator, says, "Like all her race, she had a skin several shades lighter than her husband's, being something of the colour of half-roasted coffee." As the women labour in the fields and are quite unclothed, it is not likely that they differ in colour from the men owing to less exposure to the weather. European women are perhaps the brighter coloured of the two sexes, as may be seen when both have been equally exposed.[70]

Darwin went no farther. He probably had trouble imagining how males could select females in the human species when females do the selecting in most other species. "Throughout the animal kingdom, when the sexes differ in external appearance, it is, with rare exceptions, the male which has been the more modified [...] Hence it is the males that fight together and sedulously display their charms before the females [...]."[71] After insemination, the male is usually free to look for another mate, while the female is tied down by pregnancy and infant care. Males thus outnumber females on the mate market and are the ones that have to make a song and dance. Females get to select, and males get selected.

That is how things are in most species, particularly mammals. Why should humans be different?

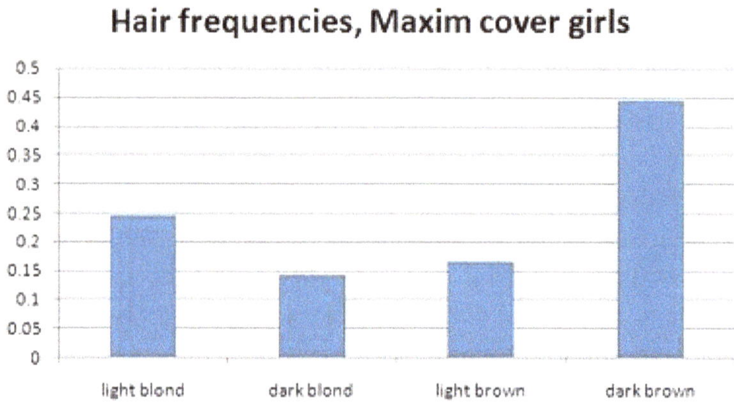

Figure 3: *Maxim* cover girls by hair color. They are much more often light blond or dark brown than dark blond or light brown, and yet the latter shades are more frequent in real life (Anon 2008b)

[70] Darwin 1936[1888], p. 867.
[71] Darwin 1936[1888], p. 578.

Sexual selection was not equally strong everywhere

Whatever he may have thought about hair, eye, and skin color, Charles Darwin did believe that sexual selection had produced visible differences among human groups: "[...] the races of man differ from each other and from their nearest allies in certain characters which are of no service to them in their daily habits of life, and which it is extremely probable would have been modified through sexual selection."[72] This happened because "the different races of man differ in their taste for the beautiful."[73]

Do they? A "taste for the beautiful" does exist. A series of experiments has shown that children as young as two to three months look longer at female faces that adults have rated as attractive. But this happens even when white infants are shown faces of black women that black men find attractive and, conversely, when black infants are shown faces of white women that white men find attractive.[74] Similar findings have been obtained with adults of various ethnic backgrounds.[75] There are some differences. East Asian men tend to prefer immature and inexpressive faces whereas African American men tend to prefer women with large buttocks and heavy body build.[76] Nonetheless, such differences in preference seem to be much smaller than the differences between the same groups in outward appearance.

Though not writing about humans, Darwin described another way in which sexual selection can produce visible group differences. If one

---

[72] Darwin 1936 [1888], p. 908.
[73] Darwin 1936 [1888], p. 888.
[74] Langlois et al. 2000; Langlois et al. 1991; Langlois et al. 1987; Langlois and Stephan 1977.
[75] Bernstein et al. 1982; Cunningham et al. 1995; Maret 1983; Miller 1969; Perrett et al. 1994.
[76] Cunningham et al. 1995.

group has undergone stronger selection than the others, it will diverge from them in appearance even if its notions of beauty are the same. It will become different-looking because its notions of beauty are more often translated into actual mate choice. Sexual selection is stronger if the group has a skew in its "operational sex ratio;" in other words, the group has a numerical imbalance between males and females among individuals available for mating. Perhaps the more numerous sex has a lower death rate or perhaps some of them are polygamous and taking more than their fair share. Either situation will favor the most attractive individuals. Their appearance will be passed on to their offspring, and the group as a whole will look increasingly different from groups with less skewed ratios. [77]

Let's apply the above model to the evolution of European hair, eye, and skin color. What conditions does the model have to satisfy? First, sexual selection must have been unusually strong among Europeans when they were still hunter-gatherers. As we will see, one part of Europe already had humans with pale skin and diverse hair and eye colors when farming first arrived.[78]

Second, studies of recent hunter-gatherers have found that the operational sex ratio is skewed either toward a male surplus by polygyny or toward a female surplus by early male death. This happens because polygyny (one man with more than one wife) is much more common than polyandry (one woman with more than one husband), and because early death happens more often to men than to women.

Third, the operational sex ratio is skewed differently in different groups of hunter-gatherers. The polygyny rate is affected by the sexual division of labor, which in turn depends on the climate and the terrain. The risk of early male death is affected by the hazards of hunting, which likewise depends on the climate and the terrain.

If we bring the above factors into our model, we see two trends as we go from lower to higher latitudes during the time when all humans, including Europeans, were still hunter-gatherers:

---

[77] Darwin 1936, pp. 573-578; Emlen and Oring 1977.
[78] Günther et al. 2018; Mittnik et al. 2018.

- at higher latitudes, a lower polygyny rate due to the increasingly limited food autonomy of women and the increasingly heavy burden on men to provide food for everyone in the family
- at higher latitudes, a higher death rate among young men due to the hazards of longer hunting distances and less hospitable terrain

In groups farther from the equator, more women had to compete for fewer men on the mate market. They were thus subjected to stronger sexual selection. We will now discuss this model in detail.

From the Tropics to the Arctic

Among tropical hunter-gatherers, women provide about half of the food for their families by gathering berries, fruits, roots, grubs, eggs, and other sessile items, such tasks being more compatible than hunting with the demands of pregnancy, breast feeding, and infant transport.[79] Because a man expends less effort overall than a woman on child care, he can more easily take an additional spouse. Indeed, he best serves his reproductive interest by having as many as possible, the number depending more on his ability to fight off rivals than on his ability to provide for an additional family.[80]

Marginal environments, like deserts, offer women less gatherable food, and men become the main providers. Such environments make it harder for a man to support a second spouse and her offspring. In the Kalahari, "having two families simultaneously is difficult to manage, both economically (in providing for a large number of dependents) and socially (in avoiding the conflicts and irritations of polygynous marriages)."[81] Each additional wife, with her offspring, adds to the burden of net food-consumers.[82]

As we go farther from the equator, we see less food gathering by women, especially in winter.[83] Below 40°N they provide 40-55% of all food consumed by hunter-gatherers. Above 60°N the proportion is less

[79] Kelly 1955, pp. 268-269.
[80] van den Berghe 1979, pp. 65-67.
[81] Howell 1979, p. 272.
[82] Howell 1979, pp. 53-54.
[83] Hoffecker 2002, p. 8; Kelly 1995, pp. 262-270.

than 10%.[84] In the Arctic, opportunities for gathering are few and far between. A study of the Caribou Inuit found very little food gathering: women gather eggs during two weeks of the summer, as well as raw gadfly larvae, the root of a plant species, and some berries.[85] For the most part, Inuit women process meat and do tasks unrelated to food procurement, such as garment making and shelter building. Men procure almost all the food.[86]

Men thus find it costly to take additional wives:

> Since the duty of being a provider was more onerous for the man who had two or more wives, this meant that as a rule it was only the really able and skilful hunters and fishers who could manage this double duty. [...] The economic conditions for an existence in the Eskimo area imposed certain restrictions on the taking of more than one wife. To be able to provide for more than one wife, as has been mentioned earlier, a man had to be a very clever hunter or sealer in the traditional Eskimo communities.[87]

> Monogamy is prevalent, as the support of several wives is possible only for the expert hunter.[88]

Monogamy likewise prevails among the Chukchi of northeastern Siberia: "The Chukchees commonly live in monogamy; it is only exceptionally that they have two wives."[89] A similar observation appears in the earliest ethnography of the Sami (Lapps) of Scandinavia: "Neither is it Lawful for them to Marry more than one Wife at a time, or to be divorced from her. Polygamy and Divorces [...], are Things unknown to the Laplanders, both whilst they were Pagans, and since."[90] The low polygyny rate at high latitudes has left its mark on the genome. Proportionately more men have contributed to the gene pool at temperate and Arctic latitudes, as shown by the higher ratio of Y to X chromosome variability among Europeans, Asians, and Amerindians in comparison to long-established

---

[84] Martin 1974, pp. 16-18.
[85] Birket-Smith 1929, p. 133.
[86] Hoffecker 2002, p. 8.
[87] Kjellstrom 1973, pp. 118-119.
[88] Hodge 1959[1905], p. 809.
[89] Nordenskiold 1882, p. 504.
[90] Scheffer 1704 p. 296.

tropical populations.[91]

The mate market is affected not only by the polygyny rate but also by the risk of early male death. Among the Kung of the Kalahari the sex ratio falls from 105 males per 100 females at birth to 91 males per 100 females at 10 years of age and over.[92] The Yanomami of Amazonia show a similar reversal between birth and adulthood.[93] Men die earlier because the sexual division of labor assigns gathering to women and hunting to men. The latter activity is riskier because a hunter covers more territory and encounters more hazards.[94] The male death rate is therefore proportional to the average hunting distance.

As we go farther from the equator, we see an increase in the male death rate. There are several reasons. First and foremost, hunting distances are longer: game animals roam over a larger territory because the cooler temperature reduces the amount of vegetation they can feed on. Second, more time is spent hunting because women cannot gather as much food, particularly in winter. Third, there are fewer alternative food sources if game animals are unavailable. Finally, the land is colder, especially at night, and may not always provide suitable materials for temporary shelter. Men are therefore at greater risk of death from starvation or exposure.

The sex ratio is highly skewed among Arctic hunter-gatherers. In the nineteenth century, the Labrador Inuit had only 57 males for every 100 females in the 15+ age bracket.[95] In all groups of Inuit, "the preponderance of adult women is generally explained by the higher death rate among men due to the natural hazards of hunting."[96] Among the Labrador Inuit,

---

[91] Dupanloup et al. 2003. See also Scozzari et al. 1997; Torroni et al. 1990.

[92] Howell 1979, p. 247.

[93] Chagnon et al. 1979.

[94] Among the !Kung, "gathering, as opposed to hunting, does not seem to be a highly risky business" (Howell 1979, p. 57). Because early death more often happens to men than to women, and because the polygyny rate is relatively low among the !Kung, there is "a relative scarcity of husbands" and most women lose some reproductive time due to involuntary celibacy (Howell 1979, pp. 247-248).

[95] Scheffel 1984.

[96] Weyer 1932, pp. 135-136. A surplus of female adults exists despite female infanticide in the northernmost parts of Inuit territory. In the High Arctic, parents view a daughter's prospects as problematic. She will have trouble finding a husband in the local group because the low carrying capacity of the land limits the group's size. She will thus probably marry into another group and be too far away to support her parents

"hunting accidents such as drowning and exposure constituted a major threat to men between approximately fifteen and fifty years of age."[97] Among the Netsilik Inuit:

> [...] drowning, either during kayak-hunting or when crossing lakes or rivers, seems to have been the most frequent type of accident. Hunters were also devoured by polar bears. Starvation, too, seems to have found more victims among men. Of the 25 deaths from hunger counted by Rasmussen, 16 were men and only 9 women. Furthermore, men had a greater tendency to commit suicide than females. Finally, in all of the murder cases described by Steenhoven the victims were men.[98]

According to a study of indigenous Siberian peoples, hunting-related deaths are due mainly to lengthy travel with a minimum of possessions. The death rate is higher for Chukchi men, who follow reindeer herds on foot, and lower for Nentsy men, who cross the tundra on sleds and use herd dogs to keep reindeer nearby.[99] The Chukchi thus resemble earlier nomadic peoples who could not control the movements of reindeer and had to follow them over long distances and across hazardous terrain. In addition, herds fluctuate in size from season to season, from year to year, and over longer cycles. Hunters of wild reindeer "are more frequently in extreme situations—of both feast and famine—than are hunters of most other faunal species."[100]

Hunting distances are longest in the continental Arctic, where hunters have to cover large areas to secure highly dispersed and mobile prey.[101] In maritime environments of the High Arctic, they cover shorter distances in pursuit of solitary game, fish, and seals.[102]

---

later on, either directly or through her future husband (Balikci 1967; Riches 1974; Smith and Smith 1994). At lower Arctic latitudes, the land has a higher carrying capacity, and female infanticide occurs much less often, apparently because bigger and closer groups enable women to marry locally and provide their parents with son-in-law payback (Riches 1974; Schrire and Steiger 1974).

[97] Scheffel 1984, p. 63.

[98] Balikci 1967, p. 623. See also Burch 1972, pp. 339-368, 347, 349-350; Kelly 1995, pp. 128-132.

[99] Krupnik 1985.

[100] Burch 1972 p. 360.

[101] Hoffecker 2002, p. 8.

[102] Kelly 1955, p. 129.

Steppe-tundra of the last ice age

Thus, as we go farther from the equator, we see that hunter-gatherer groups have fewer polygynous men and fewer men altogether. The result is a surplus of unmated women.

The surplus is largest in an environment that no longer exists: the steppe-tundra that covered the plains of northern Eurasia during successive ice ages until the end of the Upper Paleolithic some ten thousand years ago (Figure 4). Unlike today's northern barrens, it enjoyed good growing conditions due to a combination of fertile loess soil and warm summers. This was especially true at its western end, where the Scandinavian icecap had pushed the steppe-tundra zone southward into sunnier latitudes and where the nearby Atlantic provided warmth and moisture.[103] Such conditions favored the growth of grasses, mosses, lichens, and low shrubs, which in turn supported large herds of reindeer, as well as bison, horses, and mammoths. At the top of the food chain was a large human population. Despite the high bioproductivity, humans had to cope with several challenges: winter temperatures averaging -20 to 30°C, with little natural protection; scarce wood for fuel or shelter; and a food supply almost entirely limited to herds of migrating herbivores.[104]

In the steppe-tundra we see a culmination of the two latitudinal trends we discussed earlier: (1) women with the lowest level of food autonomy and (2) men with the highest risk of early death. First, women depended almost wholly on men for food, in the form of meat. A meat diet is indicated by the volume of animal bone at Upper Paleolithic sites, by the lack of wear on human dentition—a sign of a grit-free, carnivorous diet,[105] and by biochemical analysis of human remains that reveals a diet high in animal protein.[106] Polygyny would have been too costly for all but the ablest hunters. Second, men hunted over long distances and ran a high

---

[103] Butzer 1964, pp. 144, 463; Hoffecker 2002, pp. 3, 17.

[104] Hoffecker 2002, pp. 7-12, 21-26, 32-34; Schild 1976.

[105] Dickson 1990, p. 180. See also Butzer 1964, p. 374; Dahlberg and Carbonnell 1961.

[106] Richards et al. 2000. Richards et al. 2001. Food gathering was not totally absent. A Czech site has yielded evidence of fleshy taproots from charred plant remains (Mason et al. 1994). The remains are dated, however, to approximately 26,000 BP, when forest-steppe predominated. Furthermore, the investigators concluded that such foods were limited to southern and central Europe.

risk of early death. Long hunting distances are indicated by the presence of large quantities of black flint at central Russian sites. The sources are at least 130 to 150 km away, and other raw materials have sources up to 650 km away.[107]

Men were therefore in short supply. It was once thought that males predominated among human remains from Upper Paleolithic Europe,[108] but reanalysis with improved sexing criteria has found a predominance of females.[109] The Maszycka Cave in Poland provides a snapshot of a single extended family: the remains of three men, five women, and eight children.[110]

The shortage of men intensified sexual selection, which in turn radically changed the outward appearance of humans. The change went farther in some groups on the Eurasian tundra than in others. In the European section, where the climate was warmer and wetter, human groups were not only larger but also more continuous over time and space. This was where sexual selection had the best chances of creating new features that would persist and accumulate.

Human groups were smaller and less continuous in north Asia. The steppe-tundra was colder and drier there, being farther north and farther from the Atlantic's moderating influence. Conditions were especially harsh at the height of the last ice age, when many groups disappeared entirely.[111] This die-off has been shown by genetic analysis of skeletal remains from south-central Siberia, dated to 24,000-17,000 years ago: the north Asians of that time had stronger genetic affinities with present-day Europeans and Amerindians than with present-day north Asians, who seem to descend largely from people who came from the south near the end of the ice age.[112] Nonetheless, a European-looking group did survive into later times in one part of north Asia. According to another genetic analysis of remains from south-central Siberia, dated to a period extending from the third millennium BC to the fourth century AD,

---

[107] Hoffecker 2002, pp. 184-185, 248.
[108] Binford 1968; Harrold 1980; Vallois 1962.
[109] Mallegni and Fabbri 1995.
[110] Kozlowski and Sachse-Kozlowska 1995.
[111] Goebel 1999; Graf 2009a; Graf 2009b.
[112] Maanasa et al. 2013.

most of the individuals had blue or green eyes, blond, red, and light brown hair, and fair skin.[113] Indeed, old Chinese records describe south Siberian peoples as having "green eyes" and "red hair."[114]

The steppe-tundra, particularly in its European section, was unique among the many environments that humans encountered as they spread into Eurasia. Food was abundant but accessible only to men of hunting age, whose ranks were thinned by hunting-related mortality. Mateable women thus outnumbered mateable men and were subjected to intense sexual selection. Today, that region still appears as a singularity on the world map of human physical variation. If we add the areas covered by the Scandinavian icecap, which were repeopled after the last ice age, we see that it roughly corresponds to the same region where the skin is whitest and the hair and the eyes most diversely and brightly colored.

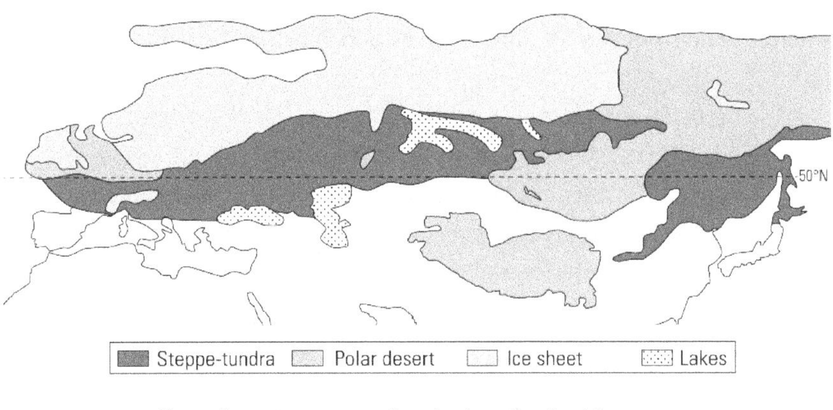

Figure 4: Eurasian steppe-tundra during the last ice age. In its western end, it was pushed south by the Scandinavian icecap into sunnier latitudes, while receiving further warmth and moisture from the nearby Atlantic.

Testimony of ancient DNA

Soft tissues begin to decay soon after death, and a corpse is normally reduced to skeletal remains after a decade in the ground. There are exceptions, but none of them takes us far enough into prehistory. Thus,

---

[113] Bouakaze et al. 2009.
[114] Keane 1886, p. 703.

in depicting how humans looked in the distant past, especially their hair, eyes, and skin, paleontologists have often had to use guesswork and imagination.

All of that is changing with the retrieval of DNA from human remains. We can now study the genes that produced long-gone soft tissues, thus putting flesh back on dead bones and revealing how people looked in earlier times. In particular, we can find out when early Europeans acquired their pale skin and their diverse hair and eye colors.

That phenotype was absent over most of Europe until late in prehistory. Instead, there was a curious combination of dark skin, dark hair, and blue eyes, as shown by DNA dated to 11,000 years ago from England, 8,000 years ago from Luxembourg, and 7,000 years ago from Spain.[115] The old phenotype persisted even later, as attested by an individual from England who lived 4,000 to 5,000 years ago and had intermediate to dark skin, brown eyes and black or possibly dark brown hair.[116]

A different picture emerges from the DNA of hunter-gatherers who lived 9,500 to 6,000 years ago in Scandinavia and 7,460 to 5,360 years ago in the East Baltic.[117] They already had pale skin and a variety of hair and eye colors. The same phenotype apparently prevailed farther east; indeed, the oldest genetic evidence of blond hair, dated to 18,000 years ago, comes from central Siberia.[118] As for pale skin, inferential methods put its time of origin at 19,000 to 11,000 years ago, according to one research team, and 19,200 to 7,600 years ago, according to another.[119]

The same period also saw changes in hair form. Hair used to be thick, straight, and black in early Europeans, as it still is among the indigenous peoples of north and east Asia. Thickness and straightness is notably due to a derived allele at the *EDAR* gene, which appeared not long after modern humans entered northern Eurasia some 30,000 years ago.[120] That allele was still present in half of Europeans as late as 8,000 years

---

[115] Brace et al. 2019; Lazaridis et al. 2014; Olalde et al. 2014.
[116] Brace et al. 2019.
[117] Günther et al. 2018; Mittnik et al. 2018.
[118] Mathieson et al. 2018.
[119] Beleza et al. 2013; Canfield et al. 2014.
[120] Kamberov et al. 2013.

ago.[121] Today, its incidence is ~1% in Europeans versus 87% in East Asians.[122] Europeans thus went from thick straight hair to thin hair with diverse forms: ~45% have straight hair, ~40% wavy hair, and ~15% curly hair.[123] The cause was probably the same desire for novelty that diversified hair and eye color. A novelty effect has in fact been shown in an Austrian study: the female participants tended to change their hair form to a less common one.[124]

It seems that Europeans acquired their current appearance during the last ice age, perhaps ten to twenty thousand years ago on the plains stretching from the Baltic to Siberia. The new phenotype would die out in its eastern range while spreading to the rest of Europe ... on the eve of recorded history. Only then did all Europeans come to look "European."[125]

---

[121] Mathieson et al. 2015.

[122] Unterländer et al. 2017.

[123] Medland et al. 2009.

[124] Schweder 1994.

[125] That demographic expansion, and the resulting contacts with different-looking people, might be dimly recalled in folklore and mythology. An ancient Norse poem, the *Rigsthula*, describes how the god Rig created a class of black-haired, swarthy, and flat-nosed thralls. The same trope appears elsewhere in Old Norse literature (Jonassen 1951; Karras 1988).

Red-haired women are special

The various hair colors differ naturally in frequency between men and women, with red hair showing the largest sex difference.[126] When the highest gradation of redness was examined in a Czech population, it was found that 19% of the women and 11% of the men had that hair color.[127] There is thus a positive interaction between being a woman and having red hair. This interaction, in turn, interacts with certain health issues. Female redheads are more sensitive to pain, whereas male redheads are like everyone else.[128] Red hair has been linked to other conditions that may be sex-linked: endometriosis, Parkinson's disease, decreased platelet function, and perhaps defects in the immune system.[129]

What do redheads and women have in common? Apparently a higher level of estrogen. During fetal development, this sex hormone seems to shift pigment production in hair follicles toward a redder mix. If you are born female and red-haired, your body tissues have a higher probability of being estrogenized beyond the normal human range. Female redheads should therefore be more prone to estrogen-related health issues.

I and two Czech researchers, Karel Kleisner and Jaroslav Flegr, tested this prediction with data from a health survey of over seven thousand people. Male redheads were as healthy as other men, doing better on average in three categories and worse in three. Female redheads, however, did worse on average than other women in ten categories and better in only three. They were especially prone to four types of cancer:

---

[126] Shekar et al. 2008; Hysi et al. 2018.

[127] Frost, Kleisner, and Flegr 2017. See also Figures 5 and 6. Redness of hair color was graded on a scale of 1 to 6 (1 = no redness, 6 = highest degree of redness).

[128] Andresen et al. 2011; Liem et al. 2005.

[129] Woodworth et al. 1995; Missmer et al. 2006; Tell-Marti et al. 2015; Gao et al. 2009; Chen et al. 2017; Liem et al. 2003; Bannatyne, Skowron, and Weber 1969.

colorectal, cervical, uterine, and ovarian. Three of these are estrogen-dependent.[130]

Could there be other explanations for the increased health risk of female redheads? Jaroslav Flegr has offered three possible ones:

Stress induced by sensitivity to solar UV. Since redheads are more prone to sunburn and skin cancer, the resulting physiological stress may predispose them to various diseases. This explanation, however, is not supported by our health survey data: the rate of skin cancer was only slightly higher among redheads than among non-redheads. In fact, the difference was not significant, perhaps because the Czech Republic is too far north for the sun to be a problem. Furthermore, exposure to solar UV is associated with a different pattern of cancer risk. Whereas red-haired women were more prone to cervical and ovarian cancer in our study, a large-scale American study on the health risks of sunlight actually showed a negative association between solar UV and cervical cancer and no significant association between it and ovarian cancer.[131]

Overprotection against solar UV. Perhaps redheads protect themselves too much from the sun and fail to produce enough vitamin D.[132] They would thus be vitamin D deficient and at risk of a wide range

---

[130] Frost, Kleisner, and Flegr 2017.

[131] Ali and Vaidya 2007; Boscoe and Schymura 2006.

[132] Flegr and Sýkorová 2019. In reality, most people with pale skin do little to protect themselves from the sun, especially at younger ages. Regular sun protection is practiced mainly by older people who see first-hand the long-term effects of sun exposure.

> Several studies have reported that although teens know that sun exposure is harmful to the skin, they continue to sunbathe and use artificial UVR tanning beds. We also found this among our participants. One survey indicated that, in spite of 78 per cent of teens knowing that sun exposure damages the skin, as many as 66 per cent nevertheless opted to tan because they believed that they look healthier with a tan. Other studies have revealed that some allow themselves to tan simply because they find covering up in the sun to be 'a hassle' (Dennis et al. 2009)

Among white Americans, paler-skinned people actually do less to protect themselves from the sun, apparently because of social pressure to be less pale: "Our population of frequent tanners showed the lowest sunscreen use (6 per cent) among 'fair'-skinned participants" (Dennis et al. 2009).

Furthermore, regular sunscreen use did not cause vitamin D deficiency in a study where participants were assigned either a broad-spectrum sunscreen or a placebo

of diseases. This explanation, however, is inconsistent with our previously noted finding: the rate of skin cancer was the same for redheads and non-redheads; in fact, redheads had a non-significantly higher rate. Solar UV seems to have penetrated the skin of both groups equally, and both should have produced the same amount of vitamin D.

Incomplete adaptation. Red hair may be associated with certain health issues because of incomplete adaptation. This hair color is one of the "new" ones and may have been the last to appear. If the various alleles for red hair are still adapting to the rest of the genome, and vice versa, there may still be adverse side effects. In other words, not enough time has passed for the evolution of modifier genes that can neutralize the side effects or for the appearance, through mutation, of new alleles that produce red hair without side effects.[133] This process of reciprocal adaptation would be all the more problematic because one of the five alleles for red hair seems to be of Neanderthal origin.[134]

Do redheads have more fun?

On the health survey, redheads of both sexes did better than other respondents in one category: number of children. Red-haired women also had more lifetime sexual partners.[135] The second finding is consistent with stereotypes in books and movies, which often assign female redheads "the role of sexy, strong, independent, sexually liberated, and sometimes dangerous individuals who buck societal trends."[136] According to a study of the American porn industry, 5.3% of the female performers had red hair, either naturally or by choice, although redheads are only 1 or 2% of the U.S. population.[137]

What is going on here? Is the average red-haired woman more interested in having sex with as many men as possible? Or is the average man more interested in having as much sex as possible with a red-haired

---

cream. Over the course of a summer, vitamin D levels did not fall below the reference range in any of the participants (Marks et al. 1995).

[133] Flegr 2010.

[134] Ding et al. 2014.

[135] Frost, Kleisner, and Flegr 2017.

[136] O Regan 2014, p. 7. See also Roach 2005.

[137] O Regan 2014, p. 7.

woman? A Czech study of 110 women (34% redheads) confirmed that being a redhead correlated positively with sexual activity and negatively with age at first sexual intercourse; however, the higher rate of sexual activity could not be explained by a higher level of sexual desire. The authors concluded that it was a "consequence of frequent attempts of potential mates to have sex with redheaded women."[138]

So it's true that redheads have more fun, at least the women. But what makes them so popular? Perhaps their rarity. The various hair colors have been "collected," so to speak, by a dynamic equilibrium that rewards novelty and penalizes banality. As a new color becomes more common in the population (because its bearers are rewarded with increased reproductive success), it loses its novelty value and eventually becomes so frequent that it is no more interesting than the other colors. In this dynamic equilibrium, redheads have nonetheless remained relatively uncommon, perhaps because excess mortality due to certain health issues has been removing their alleles from the gene pool. So red hair is still rather unusual and has not fully lost its novelty value.

Geographic variation in red hair

Red hair is more common in some European populations than in others. The variation is considerable even within northern and eastern Europe, with the frequency of red hair ranging from 0.2% in Germany to 8.44% in Great Britain.[139] High frequencies have also been reported from the Czech Republic. Red hair is rare or nonexistent among Finno-Ugric peoples, with the exception of the Udmurts. They have a frequency of 4% on average and over 7% in one group.[140]

It is difficult to reconcile this geographic pattern with selection by the natural environment. In all likelihood, some cultures have been more conducive than others to the survival of redheads. There was, for instance, the ancient belief that redheads were conceived in menstrual blood or through forbidden intercourse.[141] There were also beliefs that witches,

---

[138] Sykorová et al. 2021.
[139] Katsara and Nothnagel 2019
[140] Heapost 2008.
[141] Read 2012, pp. 9-10.

trolls, and Judas had red hair.[142] Whatever the rationale, discrimination against redheads has existed and has probably varied in intensity from onc culture to another.

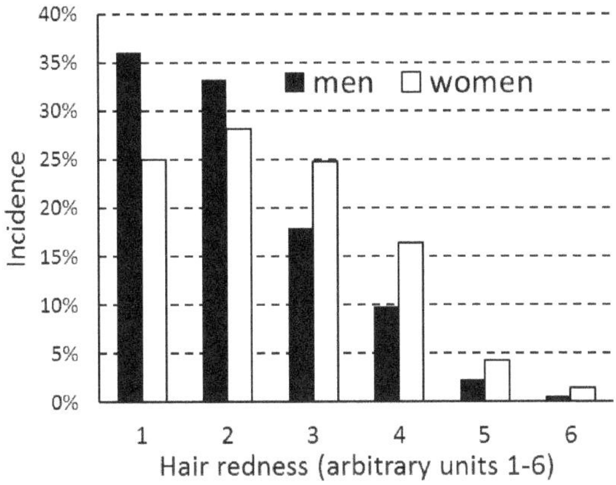

Figure 5: Frequency of red hair among Czechs, by degree of redness and by sex (Frost, Kleisner, and Flegr 2017)

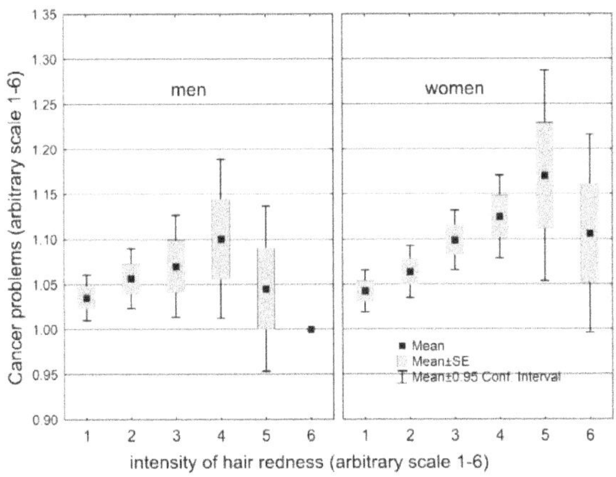

Figure 6: Relationship between degree of hair redness and seriousness of cancer. Hair redness may be a proxy, here, for estrogenization of body tissues during fetal development, the last gradation of hair redness being perhaps an exception to this trend (Frost, Kleisner, and Flegr 2017)

[142] Wimberly 1929.

Survival of the cutest? Blond hair in Oceania

After people of European descent, blond hair is most common among the natives of Oceania, within a zone stretching from central and western Australia, through Papua New Guinea, and into Melanesia and nearby islands. It is due to a *TYRP1* allele that occurs almost nowhere else.[143] This kind of blond hair coexists with dark skin and dark eyes, unlike the case with Europeans.

As with Europeans, blondism varies by age and sex. The Aborigines of central Australia are about 85% fair-haired and 15% dark-haired up to ten years of age in both sexes. Male hair then darkens, its color ranging from medium brown to black during adolescence and being almost always dark beyond the age of twenty. In contrast, female hair darkens only after twenty and is seldom darker than light brown even in old age.[144] We see similar age and sex differences in Solomon Islanders.[145] Thus, in both regions, blond hair is widespread in children of both sexes and then becomes much more specific to women. This is similar to the European pattern, but more exaggerated.

Why did blond hair evolve in Oceania? One reason may simply be an aesthetic preference:

> In Samoa, all fair hair is considered *'ena'ena*, a word that is usually translated as brown, although when English-speaking Samoans use this term in reference to hair, they typically gloss it as 'blond.' This makes sense since, when one is bleaching Polynesian hair, it goes through a series of reddish-brown shades prior to arriving at blond, and even then retains a reddish hue. When describing hair, Samoans specify the actual shade of 'blond' by using certain modifiers with *'ena'ena*, such as

143 Kenny et al. 2012.

144 Abbie and Adey 1953.

145 Norton et al. 2006.

> '*ena'ena manaia*, which literally means 'really nice brown hair,' but which refers to a very fair reddish colour.
>
> The hair of female spirits is most commonly said to be '*ena'ena manaia*, and they are wont to decorate it with a red hibiscus.[146]

Blond hair may also protect children from neglect. Papua New Guinea has a relatively high rate of child malnutrition—35% on average and up to 78% in some regions, including regions where adults are well fed. The key factor seems to be neither environmental nor ecological but cultural.[147] In some cultures of Papua New Guinea, mothers may deliberately withhold parental care:

> These women take a guarded attitude toward infants, extending the greatest amount of their affection and parental care toward children who are physically strong and who survive the first couple of years of life.[148]

Because a newborn child has uncertain chances of survival, it remains unnamed for a few weeks after birth, and the father's kin will be ritually thanked only after a six-month wait. Much will depend on the mother's feelings for the child, and those feelings may select for "cute" features, like blond hair.

---

[146] Mageo 1994.
[147] Lepowsky 1987, p. 75.
[148] Lepowsky 1987, p. 78.

Aftermath

Mesolithic

The ice age came to an end when temperatures rose to levels higher than today's. As the steppe-tundra gave way to groves of pine and birch or temperate grassland, the herds of reindeer moved farther and farther north, and hunting bands had to find other food sources. Some turned to the forest and grassland, gathering wild plant foods and hunting dispersed game like roe deer, moose, wild cattle, and wild boar.[149]Others turned to the sea, harvesting fish, seals, and shellfish.[150] They split into two groups: one adapting to the continental environment of the plains to the east and the other to the maritime environment of the Baltic and the North Sea.

Those Europeans were alone in having light skin with diverse hair and eye colors. Their neighbors to the west and south had a strange mix of dark skin, dark hair, and blue eyes.[151] Farther south, faces and bodies were distinctly African, as shown by numerous remains found in an arc stretching from Brittany, through Switzerland and northern Italy, and into the Balkans. Most are from the Neolithic, but some are as recent as the Bronze Age and the early Iron Age. Some are also accompanied by female statuettes with big breasts, protruding bellies, full hips, and large buttocks. Head hair on the statuettes looks short and matted, much like the peppercorn hair of Khoisan peoples in southern Africa.[152]

The Middle East was home to the Natufians—hunter-gatherers who were anatomically equidistant between present-day West Africans

---

[149] Schild 1976. Because game animals were less numerous, meat contributed much less to caloric intake, the result being what some authors call the "Mesolithic crisis." See Fornaciari and Mallegni 1987; Kashuba 2012, p. 4.

[150] Price 1991.

[151] Brace et al. 2019; Lazaridis et al. 2014; Olalde and Lalueza-Fox 2015; Olalde et al. 2014.

[152] Boule and Vallois 1957, pp. 291-292, 311.

and present-day Middle Easterners.[153] Yet they had more genetic affinities with the latter.[154] This should be no surprise. Changes to outward appearance involve changes to a small subset of the genome and do not reflect overall genomic change. Although Europeans and Natufians initially looked like each other and were both part of the same expansion out of Africa, the latter retained their African appearance while the former eventually lost it through changes at relatively few gene loci.

Neolithic

The Neolithic brought farming to Europe through a movement that began in the southeast and spread north and west along a broad front. The earliest farmers lived in Anatolia and Macedonia, and their remains show African nasal and facial traits.[155] They also had dark skin and brown eyes, according to DNA from the remains of three farmers who lived in northern Greece 7,500 to 5,500 years ago.[156] They were thus physically similar to the Natufian hunter-gatherers, an indication that a native group of the eastern Mediterranean had made the transition to farming on their own.

Later, those early farmers were replaced by farmers with light skin, dark hair, and dark eyes. It is unclear where the latter came from. Perhaps their ancestors had split away from the hunters of the steppe-tundra at a time when that population already had light skin but did not yet possess diverse hair and eye colors. For now, we can only say that the farmers entering Europe were not a uniform group, no more than were the native hunter-gatherers.

Did Middle Eastern farmers replace European hunter-gatherers?

DNA differs substantially between late hunter-gatherers and early farmers. That difference is often ascribed to population replacement: farmers moved into Europe from Anatolia and largely replaced the native hunter-gatherers. In the first study on the subject, the authors concluded: “these analyses provide persuasive evidence that the first farmers were not

[153] Angel 1972; Brace et al. 2006.
[154] Lazaridis et al. 2016.
[155] Angel 1972.
[156] Anon 2015.

the descendants of local hunter-gatherers but immigrated into central Europe at the onset of the Neolithic."[157] According to another study, Anatolian farmer admixture is 95% in Sardinians, 52% in northwest Europeans, 31-41% in Swedes, and 11% in Russians.[158] Those findings are cited by geneticist David Reich in his recent book: "The extraordinary fact that emerges from ancient DNA is that just five thousand years ago, the people who are now the primary ancestors of all extant northern Europeans had not yet arrived."[159]

Reich's view seems to be supported by the loss of haplogroup U, a group of mitochondrial genes that was widespread among Mesolithic hunter-gatherers across Europe and is now common only among the Sami of Finland and the Mansi of northwestern Siberia, both of whom lived from hunting, gathering, and fishing until recently.[160] Indeed, according to ancient mtDNA from central and western Europe, the population frequency of haplogroup U shows a sharp break at the time boundary between late hunter-gatherers and early farmers.[161] That break strongly suggests that the original Europeans were largely replaced by farmers spreading into Europe from the Middle East.

Yet things are not always what they seem. In Denmark, haplogroup U persisted at high frequencies long after the transition to farming, in fact as late as the Early Iron Age.[162] In Latvia and Ukraine it persisted into Neolithic times.[163] In both cases, it may have disappeared from the gene pool simply because it was no longer adaptive in a farming environment. Haplogroup U causes the energy balance to shift away from ATP synthesis and toward production of body heat—a useful cold adaptation for hunter-gatherers, who had to sleep in makeshift shelters and pursue game in all kinds of weather.[164] In contrast, farmers slept in a warmer environment and could more easily plan their outdoor activities.

---

[157] Bramanti et al. 2009.
[158] Skoglund et al. 2012.
[159] Reich 2018, p. 107.
[160] Derbeneva et al. 2002.
[161] Bramanti et al. 2009.
[162] Melchior et al. 2010.
[163] Jones et al. 2017.
[164] Balloux et al. 2009; Montiel-Sosa et al. 2006.

Haplogroup U was likely eliminated through a mix of population replacement and natural selection, the latter being more important in northern Europe. The sharp genetic break seems confined to central and western Europe, where incoming farmers advanced rapidly through territory inhabited by small nomadic bands. The wave of advance then stalled halfway along a line running from the Low Countries to the Black Sea. To the north were large semi-sedentary communities of hunter-fisher-gatherers who could less easily be replaced because they were so numerous. Their large numbers were supported by abundant marine resources of fish, seals, and shellfish.[165]

Of course, the loss of haplogroup U was not the only genetic change across the Mesolithic-Neolithic divide. Were the other changes due to natives being replaced by farmers from the Middle East? Or was natural selection again responsible? Researchers have tried to exclude the second cause by examining how noncoding genes changed across the divide, on the assumption that such genes are generally non-functional and do not affect one's chances of survival and reproduction.

That assumption is unfounded. About 80% of our genome has some kind of function, even noncoding genes.[166] Indeed, such genes may have disproportionately contributed to human evolution. Comparison of our genome with other primate genomes has shown that almost all human-specific deletions are in noncoding regions.[167] Furthermore, DNA is mostly noncoding in human accelerated regions (HARs)—genomic regions that have been well conserved throughout vertebrate evolution but are strikingly different in humans, perhaps in ways that alter how coding genes regulate each other.[168] This would be consistent with the belief that our ancestors evolved largely through new ways of regulating old systems, particularly the pace and timing of development.[169]

Clearly, natural selection caused some genetic changes during the transition from hunting and gathering to farming. One was the shift in energy balance, but there were others. Farmers had less need for odor

---

[165] Price 1991.
[166] The ENCODE Project Consortium 2012.
[167] Bae et al. 2015.
[168] Bae et al. 2015.
[169] King and Wilson 1975.

recognition, monotony avoidance, and sensation seeking.[170] They also had to process reciprocal obligations with a larger number of people while interacting less, on average, with each person. In short, farming imposed a different set of demands on mind and body; going from one way of life to the other required many physiological adjustments. Finally, there were also founder effects. When bands of hunter-gatherers are given the opportunity, only a few choose to become farmers. Because that minority is a small sample of the hunter-gatherer gene pool, the new farming population will differ genetically from the previous one in many random ways.

When reconstructing the past, particularly the transition from hunting and gathering to farming, we should not interpret genetic change solely in terms of one population replacing another. Some of the change was also due to a new regime of natural selection, as well as founder effects.

Mesolithic-Neolithic divide in the East Baltic

The population replacement model is an especially poor fit for the East Baltic. Two research teams have examined ancient DNA from that region's Mesolithic hunter-gatherers and Neolithic farmers. The first team found no evidence of Anatolian admixture in the Neolithic samples and concluded that "[t]he emergence of Neolithic features in the absence of immigration by Anatolian farmers highlights the roles of horizontal cultural transmission and potentially independent innovation during the Neolithic transition." The first signs of admixture appeared much later, in the Bronze Age.[171]

The second team similarly found no evidence of Anatolian admixture in Neolithic samples from the East Baltic. They ascribed signs of admixture in present-day DNA to limited gene flow after the Bronze Age. They also found what may be called "pseudo-admixture" in earlier hunter-gatherers. One individual had a mitochondrial haplogroup

---

[170] Majid and Kruspe 2018; Zuckerman 2008.
[171] Jones et al. 2017.

normally associated with the Neolithic expansion into Europe, while showing no evidence of Neolithic farmer ancestry.[172]

Some characteristics of the Anatolian gene pool were thus already present in some European hunter-gatherers. Other characteristics appeared much later, a pattern consistent with gradual genetic change through adaptation to a new cultural environment.

Does the Mesolithic-Neolithic divide explain current hair, eye and skin color?

It was initially thought that light skin came to Europe with the Anatolian farmers and that the current appearance of Europeans took shape during the Neolithic or even the post-Neolithic. That thinking seemed confirmed by ancient DNA from Mesolithic sites in western Europe: those individuals still had dark skin, dark hair and blue eyes.[173] Recently, however, ancient DNA has been retrieved and analyzed from Mesolithic sites in Scandinavia and the East Baltic: those individuals already had light skin and a variety of hair and eye colors.[174]

The population replacement model is problematic in another way. If Anatolian farmers had entered northern Europe some 6,000 years ago, the resulting admixture (30 to 50%) would have greatly reduced the frequencies of blue eyes, blond hair, and red hair in the native population. To bring those frequencies back to their former values, there would remain only 3,000 years before the earliest historical mentions of European hair and eye colors. Possible? Only if we assume very strong selection.

Conclusion

Were European hunter-gatherers replaced by farmers from Anatolia? A key piece of evidence is the sharp genetic break between late hunter-gatherers and early farmers. That break has been found, however, only in ancient DNA from central and western Europe, where farming rapidly replaced hunting and gathering. In Denmark and the East Baltic, change was much more gradual, both culturally and genetically.

---

[172] Mittnik et al. 2018.

[173] Brace et al. 2019; Olalde et al. 2014; Olalde and Lalueza-Fox 2015.

[174] Günther et al. 2018; Mittnik et al. 2018. See Table 4.

Some of the confusion may come from the assumption that hunter-gatherers were the same across Europe. In fact, there were regional differences in genetic characteristics, with some regional groups already sharing certain characteristics with Anatolian farmers. There were also regional differences in population density. Farmers easily replaced the hunter-gatherers of central Europe, who were few in number, but had much more trouble replacing the hunter-fisher-gatherers of Scandinavia and the East Baltic, who were already numerous and could thus hold back the advance of farming for a millennium and a half. During that time they obtained tools and weapons from farmers to the south and gradually became familiar with farming. "Agriculture became the primary subsistence regime only 1,000 years later following a period of experimentation and major changes in subsistence, settlement, and burial practices."[175] Conditions were more conducive to cultural transmission than to population replacement.

To sum up:

- A wave of farmers did spread out of Anatolia and into Europe, but it became progressively diluted through intermixture with European hunter-gatherers. Intermixture occurred especially during the long period when the wave of advance stalled along a line running from the Low Countries to the Black Sea. The last push was much more a cultural change than a demographic one.

- Although population replacement did occur during the Mesolithic-Neolithic transition, its magnitude is inflated by a model that ascribes all genetic change to that one cause. In reality, there were at least two other causes: natural selection and founder effects.

- Natural selection was important. Hunter-gatherers and farmers had to adapt to differences in diet, type of shelter, means of subsistence, and degree of sedentary living. To go from one way of life to the other required cultural adaptation and, to some extent, genetic adaptation.

---

[175] Price 1991.

Table 4. Mesolithic hunter-gatherers in Western Europe and Scandinavia/East Baltic

	Region	
	Western Europe	Scandinavia/East Baltic
Social structure	Small nomadic bands, low population density	Large semi-sedentary coastal settlements, exploitation of marine resources in summer, dispersal inland for hunting in winter
Color traits	Dark skin, dark hair, blue eyes	Light skin, diverse hair colors (blond, red, brown), diverse eye colors (blue, green, brown)
Dating of color traits (based on aDNA)	11,000 – 7,000 years ago	9,500 – 5,360 years ago
Transition to farming	7,500 – 5,500 years ago	6,000 – 3,000 years ago

Cultural legacies of an earlier time?

If the European steppe-tundra had women in excess supply, the unmated ones may have ended up specializing in activities unrelated to family life, particularly communal roles like shamanism, upkeep of camp dwellings, and tending of camp fires. That niche would have eventually become ritualized, with virginity being viewed as a mark of membership. Even when that niche no longer existed, in postglacial times, some of its ritualized aspects could have survived in daily life, perhaps into the time of recorded history.

That cultural legacy may explain a tendency of ancient European societies to associate shamanism with women, especially virgins. Old Greco-Roman, Germanic, and Slavic sources attest to an overwhelming preponderance of women among seers, witches, sibyls, oracles, and soothsayers.[176] Virgin females in particular were seen as storehouses of energy and endowed with extraordinary power.[177] Pomponius Mela (d. 45 CE) mentioned nine virgin priestesses on an island off Brittany who knew the future and gave oracular responses to sailors.[178] Cornelius Tacitus (d. 120 CE) described a virgin prophetess in present-day Germany, saying that her tribe "regards many women as endowed with prophetic powers and, as the superstition grows, attributes divinity to them" (*Histories* 4.61). The Gauls had similar prophetesses.[179] Outside Europe, that social role was not performed by single women. Siberian shamans were mostly married women.[180] Inuit and Amerindian shamans were mostly married men.[181]

---

[176] Baroja 1964, pp. 24-57.
[177] Gimbutas 1982; Dexter 1985.
[178] Chadwick 1966, p. 79.
[179] Chadwick 1966, pp. 80-81.
[180] Czaplicka 1969, pp. 243-255.
[181] Hallowell 1971, pp. 19-22; Saladin d'Anglure 1988.

Single women also performed certain communal rituals. Strabo (d. 24 CE) described a community of women on an island in the Loire where "no man sets foot" (*Geography* 4.4.6). They would unroof their temple and roof it again before sunset, a rite that may recall a time when the housewife daily removed the thatched roof to air the smoke-filled interior.[182]

Tending of communal fires was likewise done by single women. The Romans and Greeks placed the hearth under the protection of a virgin goddess, *Vesta* or *Hestia*, and a caste of virgin women, the Vestals, tended Rome's sacred fire. The word *Vesta* seems to have come down unchanged from the time of proto-Indo-European some six thousand years ago, a sign of ritualization at an early date.[183] Ancient Greeks respected the virginity of the burning hearth; in particular, men were not supposed to expose their genitals within sight of it, especially after sex.[184] That mental association has survived in some folktales, the best-known one being the story of Cinderella, a maiden who would sleep by the cinders of her stepmother's fireplace.

---

[182] Lefèvre 1900, p. 93.
[183] Dumézil 1970, pp. 311-324.
[184] West 1996, p. 337.

The original meaning of skin color

This book began with a focus on hair and eye color. Those are the two traits that most convincingly show sexual selection of European women. The focus will now move to skin color. Selection for fair skin went unusually far in Europeans, pushing depigmentation almost to the biological limit. It also went farther in women than in men, not only among Europeans but also among all humans.

Women are universally the fair sex. They are paler than men, who conversely are ruddier and browner. Today, that sex difference seems hardly noticeable in Western societies, having been overwhelmed by much larger differences of race and ethnicity and further obscured since the 1920s by the tanning fad.[185] But it was noticed in earlier times. Wherever the visual arts developed—ancient Egypt, the Greco-Roman world, early South and East Asia, Mesoamerica—female figures were given a lighter hue and male figures a darker hue.[186] The same convention appears in ancient Greek poetry: women were described as "white" and men as "black."[187]

Whiteness is still part of female identity in many non-Western societies. This was the opinion of Japanese men when interviewed on the subject: "Whiteness is a symbol of women, distinguishing them from men." "One's mother-image is white."[188]

---

[185] Segrave 2005.

[186] Capart 1905, pp. 26-27; Eaverly 2013; Pallottino 1952, pp. 34, 45, 73, 76-77, 87, 93, 95, 105, 107, 115;
Siepe 2004; Soustelle 1970, p. 130; Tegner 1992; Wagatsuma 1967.

[187] Irwin 1974, pp. 121, 129-155. This relative way of speaking corresponds to "fair" and "dark" in modern language.

[188] Wagatsuma 1967, pp. 417-418.

The sex difference in skin color: its pigments, its age changes, its hormonal causation

Skin color was first measured objectively in the 1930s, when spectrophotometers became commercially available. By measuring how much light the skin reflects across the visible spectrum, and conversely how much it absorbs, one could identify its pigments and quantify their relative importance. An American team concluded that male and female complexions differ because of differing concentrations of three pigments: melanin, hemoglobin, and carotene.[189] These differences are reduced, but not eliminated, if men are castrated and women ovariectomized. The sex hormones seem to alter skin pigmentation not only through ongoing transient effects but also through permanent organizational effects before birth and at puberty.[190]

At birth, both sexes are pale. The skin then darkens until just before puberty, but more so in girls, who actually become darker than boys. During adolescence the skin lightens, but much more so in girls. The resulting sex difference stabilizes in the late teens and persists into adulthood.[191] In addition to varying by age, this sex difference also varies by population, being larger in medium-colored groups and smaller in very light or very dark groups.[192] Finally, the sexes differ not only in

---

[189] Edwards and Duntley 1939. This sex difference is qualitatively different in sub-Saharan Africans, as shown by a study with Czech and Cameroonian participants (Fiala et al., 2022). Among Czechs, female skin reflects more light because it has less melanin and less hemoglobin. Among Cameroonians, female skin reflects more light because it has less melanin and *more* hemoglobin. An increase in ruddiness seems to make light skin darker and dark skin lighter. This may be why lighter skin is often called "red" in sub-Saharan Africa.

[190] Edwards and Duntley 1949; Edwards et al. 1941.

[191] This literature is reviewed in van den Berghe and Frost (1986). The best controlled studies are those by Kalla (1973) and Kalla and Tiwari (1970) on South Asian and Tibetan participants and Mesa (1983) on Spanish participants. After middle age, the sex difference in skin color tends to disappear:

> After middle life, the amount of estrogenic and androgenic hormones begins to diminish, and the skin of old age begins to appear. [...] there is deposited in the epidermis more pigment than in youth, and the skin becomes diffusely brown, and in those who have little natural pigment it becomes grey and pasty. In women, hair grows on the face, and finally male and female skin are nearly identical in essential characteristics. (Bancroft 1944, p.61)

[192] Frost 2007; Madrigal and Kelly 2007.

constitutive pigmentation (color of untanned skin) but also in facultative pigmentation (capacity for tanning). Men will tan more than women even when both are equally exposed to the sun. This was shown in a New Guinea study that used three variables: reflectance of skin on the unexposed upper inner arm; reflectance of skin on the exposed forearm; and time spent in the sun. Despite identical sun exposure, the men were darker than the women, and more so on exposed skin.[193] The same finding appears in another New Guinea study, whose author ruled out the possibility of the women being less exposed to the sun, "as in most parts of New Guinea the adult females are responsible for most of the food cultivation and are therefore exposed almost continuously to sunlight."[194]

The above sex and age differences are due to the action of the sex hormones, particularly at puberty. As already noted, this has been shown by studies of normal, castrated, and ovariectomized individuals. Castrated men have low androgen levels and are consequently very pale:

> One of the outstanding characteristics of a human male castrate is the paleness of the skin. After treatment with androgenic hormone, however, the individual takes on a darker and more ruddy hue. This observation suggests that the skin of the castrate is deficient in melanin and blood, and that the androgenic hormone increases the content of these substances in the integument.[195]

Estrogen has similar but much weaker effects, which are further reduced by the other female hormone, progesterone. Ovariectomy thus has much less impact on female skin than castration has on male skin:

> Ovariectomy did not remove those pigmentary characteristics which distinguish the female from the male. As previously reported, these are as follows: The female skin shows less melanin and less hemoglobin than the male. Areas primarily rich in melanin show a comparatively higher melanin ratio over poorly pigmented areas than is the case in the male. The buttocks are predominantly arterial (high in oxyhemoglobin) in the female, while they are predominantly venous (high in reduced hemoglobin) in the male. There is more carotene in the

---

[193] Harvey 1985.

[194] Walsh 1964.

[195] Edwards et al. 1941.

> female breast, abdomen, and buttocks than in the corresponding regions in the male.[196]

The action of the sex hormones has also been shown by a study of monozygotic and dizygotic twins at different stages of adolescence. The twins were divided into three age groups: 12 years, 13 to 15 years, and 16 to 18 years. It was in the second group that skin reflectance diverged the most between boys and girls, yet mean within-pair variance did not differ significantly from one group to the next, a sign of tight genetic control over age changes.[197]

Finally, a hormonal cause has been shown by two "digit ratio" studies. The digit ratio (i.e., length of the index finger divided by length of the ring finger) provides a measure of the ratio of estrogens to androgens in body tissues during development. The higher the ratio, the more the tissues have been feminized. The lower the ratio, the more they have been masculinized. The first of those two studies was done on adults. Digit ratio was found to correlate with lightness of skin in women but not in men. The correlation was also stronger for the left hand.[198] We know that the left-hand ratio is a better measure of hormonal exposure later in life, as shown by the findings of a longitudinal study of children: digit ratio increases with age, and more so for the left hand than for the right hand.[199] Conversely, the right-hand ratio is a better measure of prenatal exposure to the sex hormones.[200] The second study was done on children just before puberty, when girls are actually darker than boys. Digit ratio was found to correlate with darkness of skin in girls, but not in boys. The correlation was significant only for the right hand.[201]

Nonetheless, it is often assumed that differences in lifestyle explain the differing complexions of men and women.[202] Perhaps girls become fairer after puberty because they have less freedom to go outside unaccompanied, as is the case in many cultures. The sex difference in skin

---

[196] Edwards and Duntley 1949.
[197] Omoto 1965.
[198] Manning, Bundred, and Mather 2004.
[199] Trivers et al. 2006.
[200] Honekopp and Watson 2010.
[201] Sitek et al. 2018.
[202] See for example Eaverly 2013 and Irwin 1974.

color should therefore disappear as women become freer in their lives. That explanation seems supported by a recent study of young adults from Ireland, Poland, Italy, and Portugal. Unlike earlier studies, it found women to be darker-skinned than men on the upper inner arm, the body site typically used to measure the color of untanned skin.[203] That site, however, is increasingly unreliable for that purpose because of the growing popularity of full body tanning, especially among young women.

A cue for gender recognition

How noticeable is the sex difference in skin color? In the best controlled studies of skin reflectance, boys and girls differ after puberty by about one and a half percentage points.[204] By comparison, northern Europeans and West Africans differ by 25 to 30 percentage points.[205]

Those measurements, however, are from the upper inner arm. The breasts and the buttocks have a much larger sex difference of 6 to 15 percentage points. The usual explanation is that women dress differently and are more likely to conceal their breasts and buttocks from the sun. Yet those parts of the body are lighter-colored in women not only because their skin has less melanin but also because it has less blood in its outer layers, a double causation that is hard to reconcile with the single cause of tanning.[206] An alternative explanation is that the skin receives more estrogen on some parts of the female body than on others. The breasts and the buttocks have a thicker layer of subcutaneous fat in women, and fatty tissues are known to contain an enzyme (aromatase) that converts an androgen (androstenedione) into an estrogen (estrone).[207] A thicker layer of fat produces more estrogen and thus feminizes the adjacent skin to a greater degree, making it softer, smoother, and fairer. This hormonal explanation is supported by the parallels between age changes in skin reflectance and age changes in the fatty layer, which becomes thinner

---

[203] Candille et al. 2012.
[204] Kalla 1983; Mesa 1983.
[205] Robins 1991, Tables 7.1, 7.2.
[206] Edwards et al. 1939; Garn et al. 1956.
[207] Siiteri and MacDonald 1973.

during childhood and thicker after puberty. Indeed, the two traits correlate with each other in adult women.[208]

Whatever its actual magnitude, the sex difference in skin color seems to interest the human mind, particularly for recognition of male and female faces. Research is ongoing, but a consensus has emerged that color is "a complex and crucial feature in facial processing. [...] Further, recent work has revealed consistent patterns of connected face and color selective cortical areas, possibly reflecting a shared overlap of visual processing between faces and color."[209]

Keep in mind that we have an innate ability to recognize faces. This ability is absent in people with prosopagnosia, who may seem normal and yet are no better at recognizing a face than any other object.[210] At the other extreme are "super-recognizers," who are as good at face recognition as prosopagnosics are bad.[211] Such hardwiring should be no surprise. If a mental task is vital and occurs repeatedly in the same sort of way, it will become hardwired to shorten response time and eliminate learning time, as well as the mistakes that come with learning.

We look at faces not only to distinguish individuals from each other but also to tell men and women apart. Some neurons specialize in recognizing male faces, others in recognizing female faces, and others in recognizing both indifferently.[212] An important cue is skin color.[213] Subjects can use it to tell whether a face is male or female, even if the image is blurred and provides no other useful information.[214] This cue has two components: hue (degree of brownness and ruddiness) and luminosity (degree of contrast between lightness of facial skin and darkness of lip/eye area). Hue is the fast channel for gender recognition. If the face is too far

---

[208] Frost 2010, pp. 117-119; Mazess 1967; Parizkova 1977.

[209] Thorstenson 2018.

[210] Farah 1996; Pascalis and Kelly 2008; Zhu et al. 2009.

[211] Russell, Duchaine, and Nakayama 2009.

[212] Baudouin and Brochard 2011; Bestelmeyer et al. 2008; Jacquet and Rhodes 2008; Little et al. 2005.

[213] Bruce and Langton 1994; Hill, Bruce, and Akamatsu 1995; Russell and Sinha 2007; Russell et al. 2006; Tarr et al. 2001; Tarr, Rossion, and Doerschner 2002.

[214] Tarr et al. 2001. Indeed, facial color is especially crucial to gender recognition if face shape is not clearly visible (Yip and Sinha 2002).

away or the lighting too dim, the mind will switch to the slower but more accurate channel of luminosity.[215]

In general, darkness is associated with men and lightness with women, as shown by a series of experiments using Dutch, Portuguese, and Turkish participants. In the first one, personal names were identified by gender faster when male names were presented in black and female names in white than with the reverse combination. In the second experiment, briefly appearing black and white blobs had to be classified by gender; the former were classified predominantly as male and the latter as female. Finally, in an eye-tracking experiment, observation was longer and fixation more frequent when a dark object was associated with a male character and a light object with a female character.[216] Similar results were produced by a word-association test with Navajo participants: the color black was viewed as more potent and masculine and the color white as more active and feminine.[217]

We therefore perceive skin color through the lens of a mental mechanism that arose for gender recognition. This may explain not only why lighter skin seems more feminine but also why women have sought to accentuate their relative pallor in a wide range of cultures and time periods, most often by avoiding the sun, by wearing protective clothing, and by applying facial powders made from white clay, lime, chalk or gypsum.[218] Cosmetics almost always serve to increase the luminosity of facial skin in relation to the darker lip and eye area. The reverse visual effect is seldom desired.[219]

---

[215] Dupuis-Roy et al. 2009; Dupuis-Roy et al. 2019; Jones et al. 2015; Nestor and Tarr 2008a; Nestor and Tarr 2008b; Tarr et al. 2001; Tarr, Rossion, and Doerschner 2002.
[216] Semin et al. 2018.
[217] Osgood 1960, p. 165.
[218] Frost 2010, pp. 120-123.
[219] Russell 2003; Russell 2009; Russell 2010.

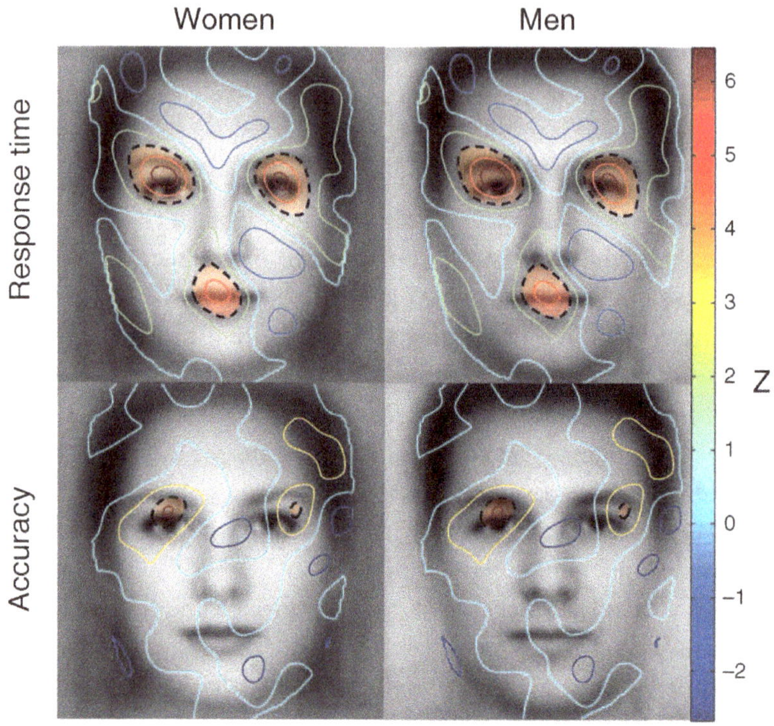

Figure 7: Averaged faces of women and men. Female faces have higher luminosity and greater contrast between facial skin color and lip/eye area color. People use this visual cue for gender recognition, particularly within the regions enclosed by the dotted lines (Dupuis-Roy et al. 2009)

A cue for sexual attraction

Facial color is a cue not only for gender recognition but also for sexual attraction. In one study, women were asked to optimize the attractiveness of facial pictures by varying the skin's darkness and ruddiness: they made the male faces darker and ruddier than the female faces.[220] In another study, a panel of mainly female participants was asked to select clothing that best matches facial skin tone: they chose 'cool' blue hues to match fair skin and 'warm' orange/red hues to match tanned skin.[221] There thus seems to be a tendency to associate redness with tanned skin and, hence, with the mix of blood and melanin that produces male

[220] Carrito et al. 2016.

[221] Perrett and Sprengelmeyer 2021.

skin tones. In a third study, the female participants associated high levels of male ruddiness with aggression, medium levels with dominance, and low levels with attractiveness.[222] Perhaps, unlike the participants of the first study, they understood the term "attractive" in a purely aesthetic sense.

Women seem to feel more attracted to darker male skin when their level of estrogen is higher. This is suggested by two studies where the female participants judged male faces at different phases of their menstrual cycle. In the first study, the women were shown pairs of facial pictures that differed slightly in color, and they had to choose the most pleasing one. When male faces were shown, the darker one was more strongly preferred by those women who were in the first two-thirds of their menstrual cycle than by those who were in the last third. During the first two-thirds of the cycle, the level of estrogen is high in relation to the level of progesterone (which acts as an anti-estrogen). During the last third, the ratio is reversed: the level of estrogen is low in relation to the level of progesterone. There was no cyclical effect among those women who were judging female faces or taking oral contraceptives.[223]

In the second study, MRI was used to measure brain activity while the female participants viewed pictures of male faces. Their brains showed a stronger response to masculinized male faces than to feminized ones, and the strength of their response correlated with the level of estrogen across their menstrual cycle.[224] In a personal communication, the lead author stated that the faces had been masculinized by making them darker and more robust in shape.

An estrogenic influence is further suggested by a third study, where preferences for darker or lighter skin were recorded for young boys and girls, as well as their body fat content. Preschoolers were presented with two dolls that differed slightly in skin color, and their adiposity was recorded by measuring their body mass index and the thickness of their subcutaneous fat. Doll choice was the same for the boys and the girls; however, among children less than three years old, those who chose the

---

[222] Stephen et al. 2012.

[223] Frost 1994b.

[224] Rupp et al. 2009.

darker doll had significantly more body fat than those who chose the lighter doll.[225] In that age range, estrogen is produced mostly in fatty tissue.[226]

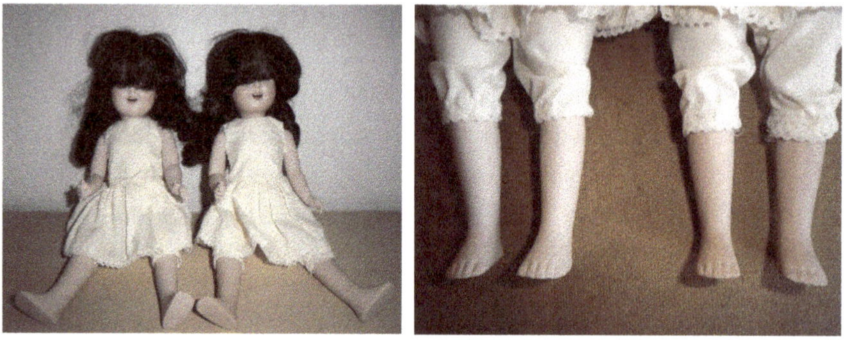

Figure 8: Two dolls that differ slightly in skin tone. Below three years of age, children who chose the darker doll had significantly more body fat than those who chose the lighter doll (Frost 1989).

A cue for behavior and emotion

Facial color is a cue for a broad range of behaviors and emotions. Because these responses have become overlain with ethnic and racial meanings, I sought to study them in a mono-ethnic context, specifically by interviewing elderly men and women in a rural French Canadian community of eastern Quebec. My interviewees described lighter skin tones in women as *faible* (weak) and *délicat* (delicate) and darker skin tones in men as *fort* (strong) and *solide* (solid). Within-gender differences were described in similar terms: a lighter-skinned person was *doux* (soft) and *facile* (easy-going), and a darker-skinned person *dur* (hard), *prompt* (quick-tempered), *orgueilleux* (arrogant), and *malin* (malicious).[227]

---

[225] Frost 1989. See also Figure 8. In other doll studies, boys and girls have similar preferences up to six years of age (Renninger and Williams 1966; Williams and Roberson 1967; Williams and Rousseau 1971). Male and female preferences begin to diverge at older ages. When a group of American children, 3 to 8 years of age, were presented with a white-faced puppet and a brown-faced one, the latter puppet was more often chosen by girls than by boys, this finding being as true for Euro-American children as for African American children (Asher and Allen 1969).

[226] Baird 1976. Body fat contains an enzyme (aromatase) that converts an androgen (androstenedione) into an estrogen (estrone).

[227] Frost 2010, pp.149-152.

These findings are in line with those from studies of color symbolism: whiteness is generally associated with weakness and blackness with strength.[228] For example, the ancient Greeks believed that a woman's light complexion incarnated her "helplessness and need of protection" and a man's dark complexion his "courage and the ability to fight well."[229] A brave and strong man had a "black rump;" a coward a "white rump." The same mental association was projected onto the internal organs and ultimately onto the soul. A "black heart" denoted strong emotions, a "white heart" indifference or a refusal to act. A coward had a "white liver."[230]

If skin tones are judged solely on the strength-weakness continuum, the darker ones would be considered more desirable than the lighter ones. If, however, we judge them on another continuum, such as threatening vs. non-threatening, we would get a different result. Perceptions of skin color are multidimensional, and we may mislead ourselves if we choose only one dimension.

This may be the case with an oft-cited finding that color prejudice is learned from parents and peers at a young age. When, in a series of studies, American preschoolers were presented with two dolls that differed in skin color, most of them chose the lighter-colored one, regardless of their racial background, when asked "Give me the doll that you like to play with" or "Give me the doll that is a nice doll."[231] Similar results were obtained with preschoolers from Japan, Italy, France, and Germany during the 1970s, when skin color had much less ethnic significance.[232] In those studies, the children associated a light-skinned figure (a person or an animal) with positive adjectives: "clean," "pretty," "kind," "helpful," "wonderful," and "good." Conversely, a dark-skinned figure was associated with negative adjectives: "naughty," "dirty," and "mean." This

---

[228] Gergen 1967, p. 397; Osgood 1960, p. 165.

[229] Irwin 1974, p. 121. See also Figure 9.

[230] Irwin 1974, pp. 129-155. In modern English this color symbolism survives in the expression "lily-livered." In ancient Greek, as in many languages, the words for "black" and "white" are often used in a relative sense to mean "dark" and "fair." Thus, Ulysses regained the "black" complexion of his youth when Athena touched him with her golden wand (*Odyssey*, xvi).

[231] Goodman 1946; Clark and Clark 1947.

[232] Best et al. 1975; Best et al. 1976; Iwawaki et al. 1978.

mental association did not follow a learning curve;[233] nor was it stronger in smarter children, who would be expected to learn it faster.[234]Later studies have reported a preference for light skin among children from Portugal, Mozambique, and Israel (light-skinned Jews and Ethiopian Jews). The preference became stronger in older Portuguese children and weaker in older Mozambican children, perhaps because in-group preferences increase with age. Israeli children showed no age effect.[235]

In sum, when the question is framed in terms of idealized feminine qualities ("choose the nice/pretty/kind/helpful doll"), children will choose the lighter-colored doll over the darker-colored one in a wide range of cultures at ages when the non-familial environment still has a weak influence. The Mozambican study is particularly interesting for three reasons: 1) the preferred facial image was much lighter than the children's own faces; 2) very light skin was preferred to fairly light skin; and 3) light-skin preference was strongest in the youngest age group (4-6 years). This preference may therefore be unrelated to race or ethnicity, perhaps being triggered by an innate maternal image in the child's mind.

That explanation is supported by two unintended findings from the above study of French children. First, when preparing the lists of positive and negative adjectives, the American researchers incorrectly used *robuste* to translate "healthy." Whereas most of the American children had associated "healthy" with the light-skinned figure, most of the French children associated *robuste* with the dark-skinned figure, perhaps because that word had masculine overtones of physical strength. Second, French being a gendered language, the adjectives were sometimes presented in the masculine form and sometimes in the feminine form. Of the 12 positive adjectives, 3 were feminine, 3 masculine, and 6 with no clear gender. Of the 12 negative adjectives, 3 were feminine, 5 masculine, and 4 with no clear gender. Of the adjectives chosen by over 70% of the children only 1 out of 5 positive ones were masculine whereas 2 out of 3 negative ones were.[236] The children thus associated light skin with

---

[233] Best et al. 1975; Munitz et al. 1987.
[234] Williams et al. 1975a; Williams et al. 1975b; Williams and Rousseau 1971.
[235] Neto and Williams 1997; Cruz 2012.
[236] Best et al. 1975.

femininity by choosing adjectives that were feminine not only semantically but also grammatically.

Figure 9: Hercules and Telephos. Fresco from Herculaneum, Museo Archeologico Nazionale (Wikicommons). A woman's light complexion incarnated her helplessness and need for protection; a man's dark complexion, his courage and ability to fight well.

Origins in the *Kindchenschema*

A woman resembles an infant in her look, feel, and voice. In comparison to an adult male, both have a lighter complexion,[237] a smaller nose and chin, smoother, more pliable skin, and a higher vocal pitch. This is what Konrad Lorenz dubbed the *Kindchenschema*, a set of visual, tactile, and auditory cues that identify a human infant to an adult, who then feels less aggressive and more willing to provide care and nurturance.[238]

A lighter skin tone identifies an infant as an infant not only in humans but also in nonhuman primates, particularly langurs, baboons, and macaques. The color is initially pink and then darkens, becoming almost

[237] Grande et al. 1994; Kalla 1973; Post et al. 1976; Visscher et al. 2017.

[238] Frost 2010, pp. 134-135; Lorenz 1971, pp. 154-164; Wickler 1973, pp. 255-265.

black in adulthood. A mother thus has a means to find her wayward offspring. Later, as their skin darkens with age, she loses interest in seeking and holding them.[239]

In humans, the infant's pink skin is especially noticeable in those societies where the adult is much darker. A new Kenyan mother may tell her neighbors to come and see her *mzungu* ("European" in Swahili).[240] Tuareg children are said to be born "white" because of the freshness and moisture of the womb.[241] When Zambian girls were asked to describe how Africans look, some wrote: "At birth African children are born like Europeans, but after a few months the color changes to the color of an African."[242] The infant coloration is often attributed to a previous spirit life:

> There is a rather widespread concept in Black Africa, according to which human beings, before "coming" into this world, dwell in heaven, where they are white. For, heaven itself is white and all the beings dwelling there are also white. Therefore the whiter a child is at birth, the more splendid it is. In other words, at that particular moment in a person's life, special importance is attached to the whiteness of his colour, which is endowed with exceptional qualities.[243]

Another Africanist makes the same point: "black is thus the color of maturity [...] White on the other hand is a sign of the before-life and the after-life: the African newborn is light-skinned and the color of mourning is white kaolin."[244]

As long as the skin remains light-colored, the newborn is soft and vulnerable, and hence at risk of returning to heaven.

> According to the same concept, it is also claimed that a newborn baby is not only white but also a soft being during the time between his birth and his acceptance into the society. Furthermore, during this entire period, he is not considered a real person, and this may go so far that parents and society may do away with him at will for reasons that are peculiar to each

---

[239] Alley 1980; Alley 2014; Blaffer-Hrdy 2000, pp. 446-448; Booth 1962; Jay 1962.
[240] Walentowitz 2008. See also Figure 10.
[241] Walentowitz 2008.
[242] Powdermaker 1956.
[243] Zahan 1974, p. 385.
[244] Maertens 1978, p. 41.

social group. Having been done away with, these beings are considered to return automatically to the place where they came from, that is, to heaven.[245]

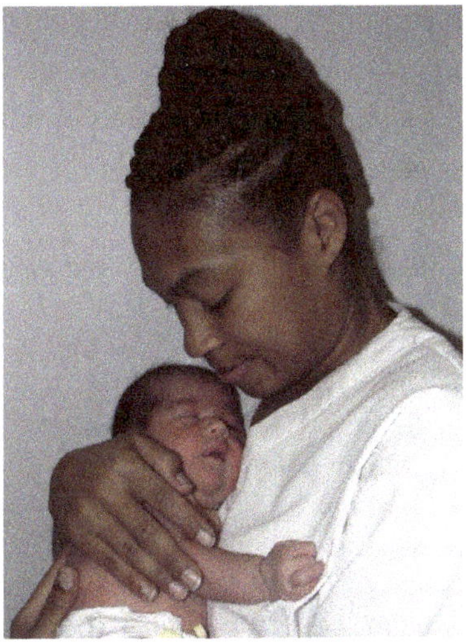

Figure 10: Newborn African American (Wikicommons)

The above evidence suggests that the skin color of infants, and later women, followed an evolutionary trajectory with four stages:

1. In nonhuman primates, the pink skin of newborn individuals initially had no mental or behavioral significance. It was literally an accident of birth, there being no need for dark pigmentation in the womb.

2. Adults recognized this visible trait as being specific to infants, and in time it was incorporated into the *Kindchenschema*. Certain mental and behavioral responses to that trait became hardwired, especially in the minds of females. This mechanism could be overridden in the minds of strange males that would sometimes invade the group and kill the young.[246]

3. Natural selection favored those infants that remained lighter-colored until they no longer had to be cared for. The infant coloration thus persists longer in those species whose infants remain helpless for a longer time.

4. In humans, a combination of slow maturation, high paternal investment, and long-lasting pair bonds made women more vulnerable to male neglect and aggression, thus putting them in a situation like that of infants and causing them to mimic the same *Kindchenschema*.

The sex difference in human skin color has no true antecedents in other primates, but there is an analogous one. In seven of the eight primate species where the male and the female differ in coat color, the female has the infant's light coloration. Interestingly, five of the seven species (63%)

---

[245] Zahan 1974, pp. 386-387.
[246] Alley 1980.

are monogamous, whereas only 18% of all primate species are.[247] By mimicking the infant's coloration, the female may improve her chances of survival in a family environment that increases her vulnerability to neglect (because she depends more on male provisioning) and to male aggression (because she cohabits longer and more continuously with the male). Indeed, such mimicry seems to have shaped much of mammalian sexuality, notably behaviors like cuddling, murmuring, nipple sucking, and mouth licking.[248]

The above evolutionary scenario was first proposed by ethologist Russell Guthrie:

> I believe the sexual differences in skin color resulted from female whiteness being selected for because it is opposite the threat coloration, although the selection pressures may have been rather mild. Light skin seems to be more paedomorphic, since individuals of all races tend to darken with age. Even in the gorilla, the most heavily pigmented of the hominoids, the young are born with very little pigment. [...] Thus, a lighter colored individual may present a less threatening, more juvenile image.[249]

By projecting a childlike image, female whiteness serves not to increase male sexual arousal but to alter it. Guthrie's hypothesis is supported by the findings of an eye-tracking study. The male participants were initially shown pictures of women and asked to rate their attractiveness: the lighter-skinned women were not rated more attractive than the darker-skinned ones. The men's eye movements were then tracked, and it was found that the women with lighter skin were viewed longer.[250] The longer duration seems to indicate not so much a higher level of sexual interest as a more sustained interest with a slower rise and fall in intensity.[251]

Interest is shown not only by length of observation but also by dilation of the pupils. In a Japanese experiment, pupil size was measured

---

[247] Blaffer-Hrdy and Hartung 1979.
[248] Wickler 1973, pp. 163-185.
[249] Guthrie 1970.
[250] Garza et al. 2016.
[251] Garza et al. 2016.

while the female participants were looking at images of female faces. There was a significant correlation between the dilation of the observer's pupils and the lightness of the observed facial color.[252] Again, the positive response to lighter skin seems to be independent of gender and is not sexual attraction per se, although it can participate in sexual attraction.

Transient changes to skin color

In women, particularly dark-complexioned brunettes, facial skin fluctuates in ruddiness and brownness, being lightest at ovulation and darkest near the end of the menstrual cycle.[253] The buttocks display an even greater cyclical variation.[254] Pierre van den Berghe and Don Symons have argued that such cyclical changes tell the male observer when a woman is most fecundable.[255] An analogy would be the sexual swellings of many female primates, that is, the vividly red area of perineal skin that swells and shrinks over the menstrual cycle and is most attractive to males when largest. On the other hand, it can be argued that the skin inevitably lightens and darkens over the menstrual cycle. The same sex hormones that produce permanent organizational effects at puberty, including changes to skin pigmentation, will necessarily produce such transient effects. There is thus no need to posit an adaptive function.[256]

Facial color can fluctuate over shorter intervals, such as during outbursts of anger. By turning red in the face, a man may seem more intimidating. When, at the 2004 Olympic Games, opponents were randomly assigned red or blue uniforms for boxing, taekwondo, Greco-Roman wrestling, and freestyle wrestling, they were significantly likelier to win if they wore red uniforms. The effect was strong: red winners outnumbered blue winners in 16 out of 21 rounds, and only 4 rounds had the reverse outcome.[257] This finding was further investigated by asking men and women to identify the color of words on a computer screen. Men

---

[252] Kuraguchi and Kanari 2021.
[253] McGuiness 1961; Snell and Turner 1966; Thorstenson 2018.
[254] Edwards and Duntley 1949.
[255] Van den Berghe and Frost 1986; Symons 1995.
[256] Nyborg 1994, pp. 32-44.
[257] Hill and Barton, 2005.

took significantly longer to answer when the words were red.[258] Other studies have shown that facial color helps an observer distinguish between emotional states that share similar facial expressions, such as anger versus disgust, surprise versus fear, and sadness versus happiness.[259]

Just as a face can turn red in anger, it can also turn pale with fright. The desired effect on the observer is now the reverse: appease and de-escalate. A questionnaire study found that pallor is associated with fear and a low level of anger. When the respondents were asked to rate their propensity for turning pale in everyday situations, the highest scores came from those who feared blood and injury. The mean score was higher for women than for men.[260]

Finally, a face can turn red with embarrassment, what we call "blushing." According to the above study, this response is not a form of appeasement and is, in fact, often associated with high levels of anger, albeit controlled anger.[261] Charles Darwin attributed it to shyness, shame, or modesty, the underlying cause being "self-attention in relation to moral conduct."[262] In most cases, the reddening is confined to the face, ears, and neck. It also seems to be heritable. Darwin cited the case of a father, mother, and ten children, "all of whom, without exception, were prone to blush to a most painful degree."[263] A series of controlled experiments has confirmed that this reddening signifies embarrassment due to a social transgression. If a person blushes while making an apology, the observer will more likely perceive it as sincere and forgive the transgression.[264]

Conclusion

Today, skin color has a primarily ethnic meaning. It had other meanings in the past, apparently even our nonhuman past. In many primate species, the infant is lighter-skinned, and this coloration not only identifies the infant as an infant but also induces feelings of caring and

---

[258] Ioan et al. 2007.
[259] Thorstenson et al. 2019a; Thorstenson et al. 2021.
[260] Drummond 1997.
[261] Drummond 1997.
[262] Darwin 1872, p. 326.
[263] Darwin 1872, p. 312.
[264] Thorstenson et al. 2019b.

protectiveness in the adult observer. At some point in human evolution, the same coloration was acquired by women, along with other visual, audible, and tactile characteristics of the infant. Skin color thus became a means to distinguish women from men, with lighter skin being unconsciously perceived as feminine and darker skin as masculine. Such perceptions would in turn influence an observer's behavioral and emotional state when interacting with a man, a woman, or a young child.

In modern Western societies, those gendered perceptions have been so eclipsed by ethnic ones that they have become largely unconscious. But they are still part of conscious experience elsewhere in the world. Perhaps this is why the subject has elicited more interest from non-Western scholars, like Hiroshi Wagatsuma in Japan and Aloke Kalla in India.

Preference for lighter-skinned women in sub-Saharan Africa

A Zambian-born sociologist, Mwizenge Tembo, was visiting his home village with his white American wife and their two boys. He lost his way and asked an elderly woman for directions. She gladly helped.

> But then she said, addressing his boys in the car, in the *Tumbuka bantu* African language:
> "*Monile asungwana, muli uli?*" ("Greetings girls, how are you?").[265]

Puzzled, he assumed the woman could not easily see the boys. Yet the same incident happened twice again, with strangers referring to the two boys as *asungwana* or "those girls." On returning to the United States, he prepared an article about his visit and chose a few accompanying photos from the hundreds he had taken of men, women, and children. Only then did the strange greetings finally make sense:

> His children, who have quite pronounced features of a Sub-Saharan African, but are light skinned compared to many relatives in the village, stood out in all the group photographs. There was nothing unusual about this obvious reality. But then he noticed something very subtle; all women had lighter glowing ambience to their skins than men although both men and women had dark skin tones.[266]

Women are lighter-skinned than men in all human populations. Just as universally, people have traditionally seen the lighter end of the local spectrum as the feminine ideal.[267] In sub-Saharan Africa, that ideal is often described as "red"—actually a reddish-brown-orange. Such a

---

[265] Tembo 2010, p. 5.
[266] Tembo 2010, p. 6.
[267] van den Berghe and Frost 1986. See Table 3 above.

color is highly prized, as Tembo noted, among the Tumbuka people of eastern Zambia:

> [...] in the *Tumbuka bantu* African language people describe women's beauty saying: "*Mwanakazi mswesi ndiye muwemi comene*" which translates as "A woman who is red-skinned is most attractive." This is more accurate than what would be the conventional translation: "A woman who is light skinned is most attractive."[268]

Similar sentiments appear in a 1950s song from the neighboring Chewa:[269]

> *Ndikonda miyendo yako*
> *Ndifiga yako*
> *Maggie ulinso*
> *Mkazi woyera*

Literally translated:

> I love your legs
> Your figure too
> Maggie you are also
> A white woman

Woyera means "white" but not in an ethnic sense. Maggie is an African woman with a naturally light complexion. This norm of female beauty is indigenous and not a legacy of European domination.

The same norm is attested in ethnographies of other sub-Saharan peoples:

> Bambara (Mali)
>
> The Bambara are not unmoved by the beauty of a woman's form; they can distinguish a well-formed body from a malformed one, a pretty woman from an ugly one, and they find a coppery skin more attractive than one of ebony black.[270]

---

[268] Tembo 2010, p. 10.
[269] Tembo 2010, p. 12.
[270] Henry 1910, p. 217.

Fulani (Senegal to Chad)

The pigmentation of the skin is an essential aesthetic criterion. It determines for the Fulani the beauty of the face. As with the Tuareg, milk bestows on the body some of its properties. It is said to give its whiteness to the dermis and also to the enamel of the teeth and to the white of the eye.

[…] The complexion of men is considered by women to be darker than their own; moreover, the aesthetic issues surrounding the epidermis are not the same. According to Fulani aesthetic norms, women ought to have a lighter skin complexion to be considered beautiful and seductive.[271]

Tallensi (Ghana)

In skin colour they vary from black through chocolate brown to bronze, which the natives call "red" (*bon-ze'e*) and regard as the most attractive bodily hue.[272]

Hausa (Nigeria)

Light skin colour, referred to as 'red,' ranks high in the Hausa criteria of beauty; many variations of colour, from black to a very light reddish brown are seen.[273]

Igbo (Nigeria)

In Ibo culture, however, these yellowish or reddish complexions are considered more beautiful than the darker, 'blacker,' complexions. [...] It is true that, in West Africa, government has for many years been identified with pale-skinned Europeans, but the Ibo evidence suggests that preference for paleness of complexion is indigenous.[274]

Azande (South Sudan)

Of the women and girls, some with babies, he kept the most beautiful in Zande eyes, those brightest of eye and clearest of skin and with full breasts, for his couch.[275]

---

271 Guilhem 2006.
272 Fortes 1945, p. 7.
273 Smith 1965, p. 264.
274 Ardener 1954, pp. 71-72.
275 Evans-Pritchard 1937, p. 60.

Berti (Sudan)

Men and women affirm without any hesitation that men are black, hot and hard and women are white, cold and soft.[276]

Somali (Somalia)

Men appreciate women of good height and stature, with good hips and breasts, and plump but not fat. A reddish tinged skin is thought highly of in preference to a dark dull black.[277]

Masai (Kenya, Tanzania)

Further requirements for being regarded as beautiful are an oval face, white teeth, black gums, a skin color as light as possible [...] [278]

Rundi (Rwanda, Burundi)

Beauty does not count very heavily, but a man is not displeased if people notice that his wife is attractive and well-fleshed, has a long and narrow nose, a light skin, and is somewhat like a cow.[279]

Ganda (Uganda)

There is, in respect of the ordinary negroid complexion, a preference for paleness deeply rooted in the Ganda ideal of beauty. [...] The Ganda concept of skin pigmentation considers light coloured complexions to be differing shades of white. A dark brown skin colour is said to be —*eruyeru*, that is, somewhat white. A really brown-reddish-yellow person is said to be *mweru* = white, which in comparison would be considered to be blonde; and this in the Ganda aesthetic language is considered as red = *myufu*, the most perfect skin pigmentation.[280]

Ila, Lunda, Luvale, and Chokwe (Zambia)

Here too words meaning literally 'white' are commonly used to refer to light skins though 'red' may also be used. Light skins

---

[276] Holy 1988, p. 471.
[277] Lewis 1962, p. 13.
[278] Merker 1910, p. 18.
[279] Albert 1963, p. 203.
[280] Lugira 1970, pp. 34-35.

are admired just as much as is shown to occur among the Ibo, and young girls discussing the possible attractions of various young men have often been heard to emphasise 'very black' as a point against someone. In the past at least one attraction of a light skin apart from its intrinsic appeal was the fact that the tattooing stood out against it in strong contrast. Very black skins are not infrequently thought to go hand in hand with inherited witchcraft and a light skin to indicate its absence. Dark-skinned women conscious of their possible disadvantage have been heard to tell men that light-skinned women will be found to be sexually unsatisfying.[281]

Ngoni (Malawi)

Young men say that what they like in a girl is a light skin colour, a pretty face, and the ability to dance and to copulate well.[282]

Kgatla (Botswana)

[...] the generally admired type is a light-skinned girl of somewhat heavy build, with prominent breasts and large, firm buttocks.[283]

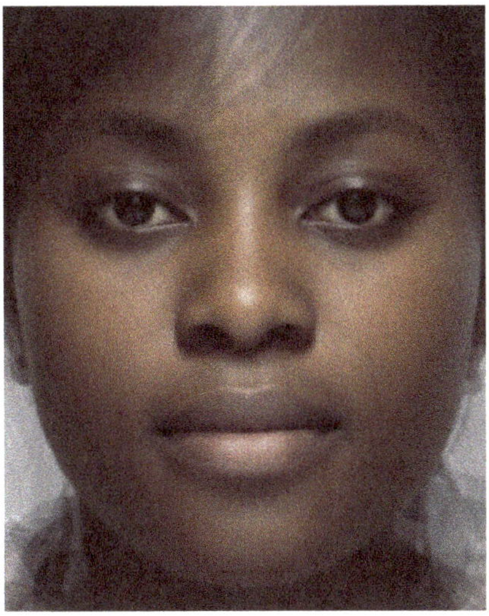

Figure 11: Composite picture of ten Cameroonian women (Credit: Karel Kleisner, Charles University). The feminine ideal is a reddish-brown-orange complexion.

An ambivalent preference

Some evidence, albeit fragmentary, predates sustained contact with Europeans. The chief of Buganda, in present-day Uganda, decided to attack the Nyambo people and seize their women … because of their brown-red skin.[284] Among the

[281] White 1954.
[282] Barnes 1951, p. 30.
[283] Schapera 1966, p. 46.
[284] Lugira 1970, p. 35.

Igbo of Nigeria, personal names from precolonial times show admiration for paler shades.[285]

Two research teams have tried to confirm this preference for lighter-skinned women in traditional African societies. The first team created standardized images of female bodies with a range of skin tones and showed them to rural men from Bakossiland, Cameroon. There was no consistent preference, perhaps because the bodies were shown only from the back with no faces visible.[286] Indeed, skin tone seems especially key to telling male and female faces apart.[287] The second research team showed facial images of real African women to urban and semi-urban men from Cameroon and Namibia. Lighter-skinned faces were preferred, but this preference became non-significant in the Namibian group when controlled for age. The younger women had lighter facial skin.[288]

The ethnographic literature likewise suggests that lighter-skinned women are viewed with some ambivalence in rural areas, where wives are chosen primarily for their ability to do farm work outdoors and in the sun.

> [...] in Nairobi of the 1960's, as throughout much of Kenya, the lighter "brown" girls are usually considered to be more beautiful than "black" girls — and the more successful prostitutes are invariably "brown." [...] Among these tribes black girls are usually regarded as hard workers, possibly because many consider themselves fortunate enough to be married.[289]

The same ambivalence has been noted among the Igbo of Nigeria:

> In the choice of a wife, yellow-skinned girls are regarded as beauties, and, other things being equal, they command higher bride prices. On the other hand it is generally held, especially by dark-complexioned persons, that yellow-skinned people are not as strong as the dark and do not live as long. A 'black' girl is said to be a harder worker. [...] A Mission headmaster was of the opinion that the preference for yellow girls was greater

[285] Ardener 1954, p. 72.
[286] Dixson et al. 2006.
[287] Dupuis-Roy et al. 2009; Dupuis-Roy et al. 2019; Jones et al. 20115; Nestor and Tarr 2008a; Nestor and Tarr 2008b; Tarr et al. 2001; Tarr, Rossion, and Doerschner 2002. See the discussion in the chapter "The original meaning of skin color."
[288] Kleisner et al. 2017.
[289] McVicar 1969, p. 242.

> nowadays than in his youth. He thought that the reason for this was that people formerly looked for strength rather than beauty and tended to marry black girls. He claimed that black people had greater powers of endurance, and he cited his own village where, he said, of the oldest six or seven people, only one was yellow.[290]

Similar opinions are expressed in other tropical farming societies where women do most of the work. In the New Guinea highlands, "the male seeks a woman of attractive physique and capability and willingness to work hard, irrespective of colour."[291]

In modern Africa, particularly in the towns and cities, this ambivalence has given way to a decided preference for women with lighter skin:

> The traditional Zulu does not make physical beauty a first priority or even an important qualification in a wife; and the skin colour of the woman is of little importance. A fair-skinned Zulu (i.e. light brown) has as much chance of winning a beauty competition as a dark-skinned one — what they call *itshi lamanzi* i.e. a river stone or *omnyama njengendoni* i.e. like a black berry. The new craze for light skins which leads to the use of face creams and other beauty preparations is a result of the Hollywood films and is typically an urban phenomenon.[292]

Imported images of beauty are indeed partly to blame. In addition, a man used to choose a wife primarily for non-aesthetic reasons: her ability to work and feed a family, her bride price, and her social background. Those reasons no longer matter as much.

Africa is changing. For one thing, there are proportionately more single women in the growing towns and cities than in the villages, urban life being less conducive to polygyny and the economic reality of supporting a second wife and her children. For another thing, images of female beauty are to be seen everywhere in advertising, fashion magazines, and social media. The African woman is surrounded by real or virtual rivals, and she responds by investing in her appearance to a greater degree.

---

[290] Ardener 1954, p. 72.
[291] Macintosh 1959, p. 281.
[292] Vilakazi 1962, pp. 59-60.

This trend is seen in postcolonial culture. During the 1980s and 1990s, Ghanaians would listen to songs describing the ideal woman as *me broni* ("my white lady") and say things like *obaa koko mbodam* ("a lighter-skinned woman doesn't lose her composure"), *obaa koko die esumu kora, a wobe hu no* ("even in the dark you can make out a lighter-skinned woman"), and *koko kata otan so* ("lightness covers ugliness"). Meanwhile, more and more women began "bleaching" their skin with cortisone or hydroquinone creams, Javel water, certain shampoos, or even cement.[293]

Today, this practice is widespread throughout Africa, particularly among the young and single. In interviews, Ghanaian women were straightforward about the reasons:

> Sometimes if you really want to marry a particular man, you have to bleach (Interview 14, 2006)
>
> Lighter-skinned women tend to attract more men by virtue of their lightness. So if they are at marrying age they get more men coming to court them earlier and quicker than darker-skinned women. (Interview 16, 2006)
>
> Darker-skinned women look at themselves and realize that they need to bleach to be beautiful. Just so men can call them beautiful (Interview 17, 2006)
>
> Even the wealthy men tend to marry the lighter-skinned women. Because when he is walking with a lighter-skinned woman... it is just beautiful. (Interview 18, 2006)

Some women would bleach to hold on to a man:

> When you talk to a lot of women, they will tell you that the reason they bleached is just like my story, my husband started cheating on me with this light-skinned girl. So I decided I needed to be light, so I started bleaching. That's what they say. I, for instance, if it wasn't for my husband cheating on me with a light-skinned girl, I would have never started bleaching, but I had to stop because I was ruining my face, especially the skin of my face. So I left him to do whatever he wants. [...] Yes when I became lighter, he stopped cheating for a little while (Interview 24, 2006).[294]

---

[293] Ondongo 1984.

[294] Fokuo 2009.

In the literature, bleaching is often blamed on self-rejection or even self-hatred—a desire to look and be European. But there is another view. While not condoning the practice, Mwizenge Tembo sees an extension of indigenous aesthetic values:

> [...] the women's desire to achieve the "*uswesi*" or orangeish-red skin tone, may have a biological basis that is embedded in the indigenous cultures among the men of the *Tumbuka*, *Bemba*, *Chewa*, and perhaps thousands of African tribes, languages, and ethnic groups. African men who may be attracted to women who have these particular skin tones should no longer have to feel guilty and defensive about charges that they harbor an inferiority complex as they have bought into the European racist belief that "white" women are the best and most attractive. Because these African women may not be trying to become "white" but may be carrying on something that was present in the indigenous African culture before Europeans established the racial ideological hegemony.[295]

That view is consistent with a Jamaican study that found no difference in self-esteem between users of bleaching creams and a control group. Motives were beauty-related:

> One participant said she loved the "brown thing" because it made her look better. Another used the bleaching cream to prevent acne and keep her skin smooth. One said she liked how it looked. Another said she did it for style because it made her look pretty. Four of the participants said they bleached because they wanted to get rid of facial pimples. The remaining participant did it because her "friends were doing it."[296]

Ironically, bleaching became popular at a time when dark-skinned peoples were reclaiming their cultural autonomy in Africa, the West Indies, and elsewhere. With more local production of videos, movies, and TV programs, local women began to see lighter-skinned versions of themselves as objects of desire. The new images of beauty have had more impact than the old ones, precisely because they are not perceived as foreign and unattainable.

---

[295] Tembo 2010, p. 17.
[296] Charles 2003.

All of this is part of a worldwide trend: a growing emphasis on beauty and an increasing array of beauty products to help women conform to their desired self-image.

Albinism

Although lighter-skinned women are traditionally admired and seen as more feminine around the world, this cross-cultural tendency has a major exception: dislike for albinos of both sexes. Albinos are especially scorned in the Tropics for their physical weakness, their shorter life span, their inability to work in the sun, and their rough, wrinkled, pockmarked skin. Among the Igbo of Nigeria:[297]

> Albinos, the extreme of paleness, however, are not admired, as their skins lack the essential smoothness and uniformity of texture and their many physical infirmities in any event render them repulsive to normal Ibo.

A similar opinion prevails among the Trobriand Islanders of Melanesia:[298]

> Albinos, with their flaxen hair and long golden body fluff, their enormous freckles, as if something dirty and brown had been splashed over them, produce an unpleasant impression on European and native alike.

In sub-Saharan Africa, this scorn can coexist, strangely enough, with a form of reverence that is nonetheless harsher in its consequences: exclusion not only from society but also from life itself, on the grounds that the albino is a heavenly being who should return from whence he came.

> In fact, he [the albino] is like heaven and remains so throughout his whole, often short life. For, in Black Africa, albinos used to be pre-eminently sacrificial victims. It was believed that their hair, mixed with seed grain, would offer the best guarantee for a good harvest. Their blood, poured over altars for crop sacrifices, was expected to do the same thing for man's food, since it would make sure that the life-giving rains would come

---

[297] Ardener 1954, p. 72.

[298] Malinowski 1932, p. 255.

> in time. How could it be otherwise, since — through the sacrifice — heaven had been introduced into the earth?[299]

Newborn children were likewise said to be light-skinned because they had just arrived from heaven.[300]

Albinos suffered less discrimination in the Amerindian societies of Mesoamerica, despite the equally strong sunlight at those latitudes and the accordingly high cost of caring for them. The Hopi of the American Southwest allowed albinos to perform in ceremonies on an equal basis and become chiefs and priests. Nonetheless, many remained single: the men because they could not work in the fields and provide for a family; the women because they voluntarily withdrew from the marriage market.[301] The same was true among the Cuna of Panama. "Moon children" were accepted in prestigious roles as healers or shamans but most never married. Those who did were usually women.[302]

And yet albinos are remarkably frequent among the Amerindians of Mesoamerica, especially those of the American Southwest. The frequency is 1 out of 200 among the Cuna, the Hopi, the Jemez, and the Zuni, versus 1 out of 20,000 Europeans.[303] The allele for albinism seems to have been passed on largely through non-conjugal relations. This mode of transmission was mentioned when Hopi villagers were interviewed about the albinos in their midst: "I knew that old man. Some say he had about twelve kids; others say about fifteen." "He never married." "They say he was always around trading with the ladies. He would make babies with them."[304] Such individuals may have enjoyed privileged opportunities to reproduce through the love and care they received.

Affection, bordering on reverence, came up repeatedly in the interviews:

> Albinos are smart, clean, nice and pretty. There is nothing wrong with them.

---

[299] Zahan 1974, p. 393.
[300] Zahan 1974, pp. 386-387.
[301] Woolf and Dukepoo 1969.
[302] Keeler 1953.
[303] Woolf and Dukepoo 1969.
[304] Woolf and Dukepoo 1969.

> Albino girls are very pretty. I like their nice color. I would like to marry one, but they probably wouldn't have me.
>
> I would like to have an albino baby. There were albinos when I was a child, so they are not new to me. In fact, I am related to one. He's my brother [clan affiliate].
>
> I know lots of them, and they are good Hopis.
>
> Here [on Second Mesa] we have lots of them and we are very proud of them. There are more here than at the other mesas. We take good care of them.
>
> It is lucky to have albinos [in the village]. They are very special.[305]

Such reverence seems to go back to pre-Columbian times. An albino may have inspired the Aztec myth of Quetzalcoátl, a god renowned for his white skin. The last Aztec emperor, Montezuma, had a special house where he "kept men and women who were white all over."[306]

In sum, the same reverence could either hurt or help the albino. It could lead to ceremonial death in one culture or to protected status in another. Above and beyond the material conditions of life, there is thus an ideological superstructure that is no less decisive. An unconscious response may be explained in different ways in different cultures, and this explanation will be ultimately integrated into a belief-system that may likewise differ from one culture to another.

Figure 12: Mem Nahadr, an albino African American performance artist (Wikicommons).

[305] Woolf and Dukepoo 1969, p. 36.
[306] Evans 2000; Woolf 1965.

The cult of feminine whiteness

Fair skin became particularly equated with femininity within a zone stretching from the Mediterranean through the Middle East and into South and East Asia. Here, pigmentation is neither very dark nor very light. It thus has the most leeway to differentiate between the sexes: female skin can lighten after puberty without running into the limit of complete whiteness, whereas male skin can darken through sun exposure without running into the limit of complete blackness.[307] Here, too, material culture first enabled women to beautify themselves by artificial means, including powders to lighten facial skin and accessories to prevent tanning—parasols, long gloves, and conical or wide-brimmed hats. Finally, here, too, appeared the first works of art, prose, and poetry, often on the theme of feminine beauty. The collective imagination thus became filled with images of milk-white women.[308]

The Middle East

Such imagery is evoked by the earliest Middle Eastern authors. A Sumerian compared his mother to "an alabaster statuette set on a lapis pedestal, a living rod of ivory, whose limbs are filled with charm." An Assyrian poet, on seeing his concubine, exclaimed: "You are white like a lizard; Your skin is excellent, like a pot."[309]

Skin color is similarly gendered in Jewish literature. In the *Song of Songs*, the bride apologizes for her dark complexion:

> Do not gaze at me because I am dark,
>   because the sun has looked upon me.
> My mother's sons were angry with me;

---

[307] Frost 2007; Madrigal and Kelly 2007.
[308] Frost 2010, pp. 35-38, 156-157. See also Figure 13.
[309] Cooper 1971; Lambert 1959.

> they made me keeper of the vineyards,
> but my own vineyard[310] I have not kept!
>
> [*Song of Songs* 1.2]

Later in time, a second-century rabbi discusses a man who vows "not to marry a particular woman who is ugly, but it turns out that the woman is beautiful; or black, but it turns out that she is white; or short but she is tall. Even if she was ugly, but became beautiful; or black, and became white."[311] A fourth-century exegetist ponders the meaning of what Abraham said to his wife upon their arrival in Egypt: "Now I know that you are a beautiful woman" [*Genesis* 12. 11]. Because the two had already lived together for some time, he interpreted the verse as meaning: "Now we are about to enter a place of ugly and dark [people]."[312] She thus looked more beautiful than ever.

Going still forward in time, we see the same gendering of skin color in Arab literature:

> [The moon] is the most common image used to represent female beauty. When attempting to draw the attention of a beautiful girl on the street, a young man may call out, "Ēš yā qamar?" (roughly, "What's happening, O moon?"). Two important components of the image, brightness (fairness of skin) and roundness (of face), convey the popular conception of beauty in Palestinian and Arab culture.[313]

> Praise of a girl's looks is traditionally couched in such terms as: Her face is like the full moon, her mouth is an almond, her nose a cardamon, she is plump, and dimpled etc. ... The highest praise is perhaps that she is as white as snow — strange praise indeed to come from a people very few of whom had ever seen snow.[314]

In his book *An Intelligent Man's Guide to the Art of Coition*, the fourteenth-century Andalusian erotologist Ibn Falita noted:

---

[310] i.e., her physical charms, including the whiteness of her skin.

[311] Goldenberg 2003, pp. 85-86. Here, the terms "black" (*shehorah*) and "white" (*levanah*) are used in a relative sense and would be better translated by "dark" and "fair."

[312] Goldenberg 2003, p. 86.

[313] Muhawi and Kanaana 1989, p. 60, see also pp. 122, 181.

[314] Haim 1978, p. 88.

> Men differ in their choice of different complexions [of women]. The majority of men prefer women of a white complexion which also has a trace of red. This complexion is said to be the best and the one that attracts the attention of poets and lovers. In fact, a smooth complexion of a pleasing colour arouses one's sexual desires. However, there are some men who prefer women of a brown complexion, which is also a pleasing colour if it has a trace of red. There are also some men who prefer black women and praise them.[315]

Figure 13: The Babylonian goddess Ishtar, Le Louvre. (Wikicommons).

The Mediterranean and Europe

On walls, vases, and statues of the classical Greco-Roman world, men were portrayed as dark-skinned and women as light-skinned: "male/female color differentiation is a widespread stylistic feature of ancient painting, one used throughout the art of the Mediterranean as well as in Egypt."[316] The same convention appears in Greek prose and poetry: women were complimented on the whiteness of their arms, faces, and feet, evidently the parts of the body most exposed to view. In the *Iliad*, Hera was described as the "white-armed goddess." To praise a man in such terms was unusual and could imply effeminacy, a dark complexion being seen as a source of strength and power.[317] Roman literature is similar.

[315] Akande 2015, p. 10.
[316] Eaverly 2013, p. 2.
[317] Irwin 1974, pp. 112-156.

With reference to the skin, the word *candor* (white) often served to praise the beauty of girls, women, and adolescent boys. It was much less common in descriptions of grown men.[318]

The cult of feminine whiteness followed the spread of civilization into Europe, building on earlier folk traditions and gaining legitimacy from them. In Old French literature:[319]

> The adjective *cler* refers to the brightness of the face, no matter whether redness, whiteness, or both are being emphasized. This characteristic is more important for women than for men. [...] The whiteness of the chest or bosom, both of which are referred to as the *piz* or *peitrine*, is emphasized in the portraits of one man and four women and is compared to that of ermine, freshly fallen snow, a lily, and a hawthorn blossom.

In Old German literature:[320]

> People liked the forehead curved and white; the nose fairly long, not broad, not hooked, but straight or a bit curved; the chin round, as white as ivory, snow or lilies and dimpled; the neck round, not too long and not too wide, of such a fine whiteness that the red wine she drinks shows through the skin. Also white are the shoulders and the nape of the neck, which tapers off softly and delicately downwards. The feminine ideal required soft white arms, round and fairly long and well-formed hands, as soft and white as ermine or snow.

In Old English literature:[321]

> The skin of beautiful women and children must be smooth, flawless, soft as silk, and above all shining white. [...] The adjective *white* is very commonly used to describe beautiful women and children and handsome men as well. The word is synonymous, for the most part, with beautiful, and when used alone, is probably meant to describe the skin in general.

Folk songs and folk tales, passed down to the present day, suggest a pre-literate tradition of idealizing women with fair skin.[322] A recurring motif

---

[318] Blümner 1892, pp. 19-20.
[319] Colby 1965, pp. 4, 59.
[320] Weinhold 1882, pp. 226-227.
[321] Curry 1916, p. 80.
[322] Child 1965, pp. (1) 86, (2) 183, 363, 399-400.
Mead 1969, pp. 321-334.

likens a perfect woman to a fortuitous combination of white, red, and black:[323]

> Once upon a time in the middle of winter when the snowflakes were falling from the sky like feathers, a queen was sitting by a window with a black ebony frame and was sewing. As she was thus sewing and looking at the snow, she stuck the needle in her finger, and three drops of blood fell into the snow. Because the red looked so pretty in the white snow, she thought to herself, "If only I had a child as white as snow, as red as blood, and as black as the wood of the window frame!" Soon thereafter she had a little daughter who was as white as snow, as red as blood, and whose hair was as black as ebony. Therefore she was called Snow-White...

This is of course the well-known fairy tale. The same motif appears in folklore throughout Europe and even in Asia and North Africa, a sign of its ancient origin and apparent resonance with the values of other cultures.[324]

South Asia

South Asian women were already using cosmetics to lighten their faces four thousand years ago: "Excavations at Harappa and Mohenjo-daro have found kohl pots and sticks for lining the eyes, as well as red iron oxides and white lead-based compounds that have been surmised to be rouge for the lips and cheeks and foundation for lightening the skin."[325] This aesthetic norm still prevails throughout the subcontinent:

> In many Indian languages the words *fair* and *beautiful* are often used synonymously. The folk literature places a high value on

---

Martinengo-Cesaresco 1886, pp. 95, 113, 117, 119, 139.
Bartok and Lord 1951, pp. 269, 279, 287, 293-297, 303, 305, 319, 357, 405, 421.
Bartok 1967, pp. 41, 45, 47, 61, 73, 83, 117, 119, 129, 149, 163-165, 167, 199, 233, 415, 513, 565.

[323] Grimm and Grimm 1969, p. 192.

[324] Cosquin 1922, pp. 1-9, 218-246. A related motif is the "snow maiden": a beautiful woman made from snow. Examples have been found in Russia and Japan (Almedingen 1957, pp. 93-105;
Seki 1963, pp. 81-82). The "swan maiden" motif, where a beautiful woman turns into a swan at will, is attested throughout Eurasia and even among the Inuit of North America (Thompson 1975, Vol. II, pp. 34, 61).

[325] Russell 2010.

> fair skin color. The ideal bride, whose beauty and virtue are praised in the songs sung at marriages, almost always has a light complexion. A dark girl is often a liability to her family because of the difficulty of arranging a marriage for her. Marriages among educated Indians are sometimes arranged through advertisements in the newspapers; even a casual examination of the matrimonial columns of such popular dailies as *The Hindu*, *The Hindustan Times*, or *The Hindustan Standard* shows that virginity and a light skin color are among the most desirable qualities in a bride.[326]

With the shift to online dating, these qualities have become even more popular, as shown by a survey of Indian matrimonial websites:

> Dark-skinned men married 'very fair' women in twelve cases, 'fair' women in 26 cases, 'wheatish' skinned women in sixteen cases, and 'dark' skinned women in only two cases. There was not one case of a dark-skinned man marrying a 'very dark'-skinned woman. Men whose skin color was graded as 'very dark,' married 'very fair' women in two cases, 'fair' women in two cases, 'wheatish' colored women in three cases, and 'dark'-skinned women in three cases. Only one case emerged where a very dark-skinned male married a very dark-skinned female.[327]

East Asia

Finally, East Asians have linked white skin to femininity since ancient times:

> In the Chinese classical tradition — as in the Japanese that so largely stemmed from it — the celebration of whiteness as a criterion of feminine beauty is a familiar theme. A poet of the fourth century B.C. celebrated a bevy of beauties for their "black-painted eyebrows and white-powdered cheeks." Of Yang Kuei-fei, the most celebrated beauty in Chinese history, the Tang poet Po Chu-i wrote: "So white her skin, so sweet her face / None could with her compare." Hands and arms of "dazzling white" move gracefully through endless reams of ancient Chinese poetry. The most common metaphor for feminine skin was white jade, and references to all the visible surfaces of jade-colored female skin abound in poets' songs. Chinese folk songs are similarly filled with the

---

[326] Béteille 1967, pp. 451–452

[327] Jha and Adelman 2009.

> whiteness of generations of beloveds: "My sweetheart is like a flower," sings one, "Please do not let the sun burn her black." In story after story Chinese writers of the 1940's were still quivering at the "snow-white" or "pure white" necks and arms of their heroines. There is some evidence that these standards have prevailed not only among effete upper-class Chinese but among rude villagers as well.[328]

The Japanese likewise:

> "White" skin has been considered an essential characteristic of feminine beauty in Japan since recorded time. An old Japanese proverb states that "white skin makes up for seven defects;" a woman's light skin causes one to overlook the absence of other desired physical features.[329]

Such women have long been sought as wives or mistresses throughout Eurasia, especially by men of wealth and power. This criterion of mate choice may explain a finding from a Japanese study on social class and skin color. Upper-class men were lighter-skinned than lower-class men, even when the latter were factory workers and not farmers and even though skin color was measured on unexposed skin. Because a wealthier man has more prospective brides to choose from, his choice will be closer to the feminine ideals of his society, including a fair complexion. For "skin color has long been regarded, by the Japanese, as one of the criteria for evaluating physical attractiveness, especially in young females."[330] Similarly, in India "[w]ealthy landowning families often have a tradition of seeking light-skinned brides among poorer members of their subcaste. It is very common to find a high concentration of lighter-skinned people among established land-owning families."[331]

Decline of feminine whiteness and rise of the tanned look

In the Western world, the cult of feminine whiteness ended during the early twentieth century, with the rapid rise of the tanned look. This fad

---

[328] Isaacs 1967, p. 370.
[329] Wagatsuma 1967, p. 407.
[330] Hulse 1967.
[331] Béteille 1967, pp. 450-451.

began with the use of sun baths and sun lamps to treat rickets, tuberculosis, and various skin disorders. Called "heliotherapy," it was developed in the 1890s by the French surgeon Antonin Poncet and the Danish scientist Niels Finsen. In 1903 the Swiss physician Dr. Auguste Rollier opened his first clinic for treatment of tuberculosis through sunbathing. The new therapy would soon be vaunted as a cure for all sorts of real or imagined ailments. "By the start of World War I, the promotion of the sun as healer was pervasive and incessant." [332]

Among women, sunbathing became so popular that a new aesthetic arose, as described by a *Los Angeles Times* writer in 1923: "Gone is the languid beauty of the winter time, with her trailing black draperies, her white face and carmined lips. Sports clothes are upon us, and with the tennis frock, the bathing suit, or the white serge tailleur goes a complexion of richest Arabian tint."[333] In the mid-decade novel *The Great Gatsby*, Miss Baker has "sun-strained eyes," a "slender golden arm, a "brown hand," a "golden shoulder," and a "face the same brown tint as the fingerless glove on her knee."[334] The new fashion may have still seemed improper in some circles: at one point, her fingers are "powdered white over their tan."[335] In 1926, a Connecticut radio station announced that "a coat of tan seems to be the latest style in natural coloring at this season of the year. [It has] been increasing in favor during the last few years."[336] In 1929, *The Delineator* informed its readers that "the glittering white sands are covered with prostrate figures all engaged in the most fashionable pastime of the day, the worship of the sun. It has become smart to look

---

[332] Segrave 2005, pp. 13-14. Graves and Hodge (1950, p. 273) saw heliotherapy as a postwar health fad of German origin. "Pacifism had been introduced from Germany at the time of the Weimar Republic. So had three other libertarian fashions — sun-bathing, nudism, and hiking. Sun-bathing had originally been found useful in Germany to cure children of 'deficiency diseases' caused by the British blockade and by the severities of the post-war years. It had now become a general cureall ..." Heliotherapy undoubtedly became much more popular after the war, but it was already widespread by 1914. The First World War, together with the Spanish flu, may have acted as a civilizational shock that accelerated changes that were already under way (Frost 2020a; McGovern 1968).

[333] Segrave 2005, p. 32.

[334] Fitzgerald 1992, pp. 15, 47, 57, 84, 185.

[335] Fitzgerald 1992, p. 122.

[336] Nickerson 1926.

healthy, smart to go in for tan, and smart to dress expressly for it."[337] Readers of *Vogue* were given similar messages: "The 1929 girl must be tanned;" "A golden tan is the index of chic."[338]

In less than a decade, sunbathing and suntanning had gone from fringe to mainstream. Its faddish acceptance was derided by a *New York Times* journalist: "Idealists would like to believe that the people, investigating the medical doctrine and accepting it as sober fact, went deliberately forth to get what was good for them, and took to sun baths with an avidity they had never shown for spinach, sleep, or orthopedic shoes."[339]

Especially avid were women. With the help of Coco Chanel and others, the tanned look would contribute to a fashion image of the 1920s: the "flapper" with bobbed hair, large shoulders, small breasts, narrow hips, and long legs. One example was Miss Baker: "a slender, small-breasted girl with an erect carriage which she accentuated by throwing her body backward at the shoulders like a young cadet."[340] Frankly androgynous, the flapper look has sometimes been ascribed to the dislocations of the First World War: either the wartime recruitment of women for previously male jobs or the postwar shortage of men. Yet the first mentions date back to 1914 in the United States, three years before that country entered the war:

> The new ideal in feminine figure, dress, and hair styles was all semi-masculine. The "1914 Girl" with her "slim hips and boy-carriage" was a "slim, boylike creature." The "new figure is Amazonian, rather than Miloan. It is boyish rather than womanly. It is strong rather than soft."[341]

The intent was to create the image of a boy on the brink of puberty, as shown by its French name: *la garçonne*.

> While the first nudist movements were organizing, *la garçonne* freely exhibited her athletic body and enjoyed the benefits of heliotherapy. The 1920s made tanning fashionable. Translucent

---

[337] Babcock 1929.
[338] *Vogue* 1929, June 22, pp. 99, 100.
[339] Segrave 2005, p. 35.
[340] Fitzgerald 1992, p. 15.
[341] McGovern 1968, pp. 324-325.

> skin and pale complexions were relegated to the theatre prop room of *fin de siècle* romanticism. Thus disappeared an element of gender differentiation. The contrast of flesh colors, brown and copper for the man, white, pink, and ivory for the woman, had been a constant since Ancient Egypt in literary and pictorial representations.[342]

The new look was further disseminated by women's magazines and by fashion designer Gabrielle "Coco" Chanel.[343] Also key was the movie industry. Joan Crawford helped spread the tanned look among Hollywood starlets to such a degree that Metro-Goldwyn-Mayer reportedly told her to stop tanning because she "looked like a lineal descendent of Sheba."[344] Nonetheless, Hollywood, like the cosmetics industry and much of popular culture, would end up promoting "colorful beauty" during the 1930s.[345]

Today, the tanned look is largely confined to the post-traditional Western world and has had trouble penetrating more traditional societies in Asia and elsewhere. This is true even in highly Westernized ones like Japan:

> [I]n the late 1960s to the mid 1970s, as soon as the fashion for tanned skin among western women came in, with images of women enjoying sunbathing on a beach in trendy places such as Nice, many Japanese women became keen on getting a tan in summer [...] but by the late 1990s Japanese women no longer unreflectively followed the western fashion. Tanned skin is still very popular and much admired among European people, whereas now very few Japanese women want to get a tan in summer. On the contrary, most Japanese women devote themselves to avoiding tanning and try to keep their skin white.[346]

[342] Bard 1998, p. 41.

[343] Coco Chanel's role in creating this fad has been exaggerated (Segrave 2005, pp. 3-11, 126).

[344] Berry 2000, pp. 110-111, 188.

[345] Weinbaum et al. 2008, pp. 48-49.

[346] Ashikari 2005, pp. 85-86.

Figure 14: Tanned skin in women's fashion. ***Vogue*****, June 22, 1929**

The dark lady of earlier times

Before the 1920s, dark female skin had likewise been frowned upon in the West, as may be seen in Victorian novels that portray "the dark lady" as an "impetuous," "ardent," and "passionate" player in short-lived romances.[347] No different are French and German novels of the same era: "[t]he love incarnated by *les brunes* appears as the conceptual equivalent of a devouring femininity, thus making them similar to the mythical figure of Lilith."[348] In his treatise on art, the French writer Pierre-Joseph Proudhon [1809-1865] described *une belle brune* "with pronounced, faintly masculine features, with these features that give a woman from eighteen to twenty-two years old satanic powers of seduction."

[347] Carpenter 1936, p. 254.
[348] Atzenhoffer 2011, p. 6.

> […] At first sight, the feeling she inspires in you is that of pity mixed with fear. One is instinctively afraid of such creatures whose passions are at times focused, at times skittish, and never quenched. There is vampire in them. Then, as you look upon *this charming head*, strangely magnetic, your pity turns into sympathy; you feel fascinated by her, seized by the demon that inhabits her. You wish, at the price of all your blood, to put out the fire that consumes her. Flee, if you value your freedom, your dignity as a man; if you do not want this Circe to make a beast of you.[349]

The French physiologist Augustin Galopin [1836-1899] claimed that men who love *les brunes* "are generally taken by a passion that is more violent, more demonstrative, more domineering, but also less profound, less tender, and less durable."[350]

The dark lady motif goes back at least to the Middle Ages and appears in the literary traditions of many European languages: the nut-brown maid in English, the *braunes* or *schwarzbraunes Mädel* in German, the *fille brune* in French, and the *barna kislány* in Hungarian. In German and Hungarian folk ballads:

> [...] dark girls [...] are inevitably imagined as sexually more available than their fairer sisters, with whom they are implicitly or explicitly contrasted. In addition, the change of a girl's complexion, such as being burned by the sun, is to be understood as symbolic of her having crossed a sexual threshold without the benefit of marriage. Doris Massny (1937) characterized the dark girl in German tradition as *verführungsbereit* (ready to be seduced). In many poems the dark girl can be a real sexual tease and, similarly, young men

[349] Proudhon 1865, pp. 245-246.

[350] Briot 2007.

Today, *une brune*, like a brunette in English, means a woman with hair darker than light brown. In the past, the meaning also included a darker complexion. Although hair color is weakly correlated with skin color, and although both traits vary continuously from one individual to another, people of earlier times tended to lump them together into opposing stereotypes: "color emigrates from the character to her clothing, from her hair to her complexion, from her complexion to her eyes, and from the character to her milieu. Two moralities are in conflict: one is affectionate and generous, represented by the *blondes*, the other demonic and mean, evoked by *les brunes*" (Atzenhoffer 2011, p. 7).

> who approach her can be free with their language in telling her what they are after.
>
> In Hungarian tradition we also frequently see the preference for the more available dark girl over the fair one, as in the following example, where a young man at first has difficulty making up his mind about what he desires. ... "The hut is ablaze, the reed crackles,/ how I love the dark one! Once I loved a blond girl,/ didn't care the least for the dark one./ Now I long for the blonde, / just like I long for grapes;/ but I long even more for a dark one,/ like I long for a tart cider apple." ... "The reed by the fence is ablaze,/ I love only the dark one;/ softly shines the moon,/ dark is certainly more beautiful than blonde" ... In a Hungarian example, the young man roundly rejects fair girls but tells us in rather vulgar terms that he plans to sample as many dark girls as he can.[351]

The same motif is echoed by sayings along the lines of "the darker the berry, the sweeter the juice," as in this Venetian folk-poem:

> "My lady mother always told me that I should never be enamoured of white roses," says a sententious young man; "she told me that I should love the little mulberries, which are sweeter than honey." "Cara mora," *mora*, or mulberry, meaning brunette, is an ordinary caressing term.[352]

In sum, dark skin could be seen as erotic in a woman, even before the tanning fad of the twentieth century, but the eroticism was more intense and shorter-lasting than in the case of fair skin. As such, it encountered social disapproval.

---

[351] Vasvari 1999; Massny 1937.

[352] Martinengo-Cesaresco 1886, p. 95. In these examples, skin color is not used to describe an exotic other. Dark and fair are simply opposing ends of the range of European pigmentation.

A forgotten past

In its heyday, the cult of feminine whiteness produced a global market for fair-skinned women as wives or concubines. Indeed, European women were once sold in large numbers to clients in Muslim Spain, North Africa, the Middle East, and South Asia. That trade, known as "white slavery" in English and *la traite des blanches* in French, began in the eighth century and would persist in some parts into the nineteenth century.

At first, the women were captives from Muslim-conquered Spain and southwestern France:

> The same convoys of booty also included women, these Frankish women who were all the more sought after in Cordova because they were blonde and fair-skinned. It was among them, as among the captive women from Gascony, that the Umayyad princes chose their most pampered concubines and who, once they became mothers, were themselves raised to the rank of veritable princesses, of proven sultanesses (*umm walad*) who were influential and quick to enter, with the assistance of Slav eunuchs, into secret and complicated palace intrigues. But the Frankish women did not populate only the caliph's harems; the dignitaries of the *khassa* and the rich burghers of the cities also procured them at lavish prices [...][353]

The trade continued into peacetime, its source of supply now being mainly the pagan lands between the Elbe and the Volga, the homeland of Slavs still unconverted to Christianity. Fighting among rival tribes or with Germans on the Elbe produced unwanted captives, often entire villages, who were kept for sale to traveling merchants:

> Adults and children were eliminated for obvious reasons. They did not correspond to the Muslim demand for young virgin girls and beardless boys and it was out of the question to gather children and raise them. The traders had neither the time nor the willingness and more importantly it would not have been cost-

[353] Lévi-Provençal 1953, p. 179.

> effective. Later, they were more willing to spare the lives of captives, by selecting them according to their capacities for productive work and by using them to the limit of their strength at laborious physical tasks.[354]

Despite condemnations by clerics like Agobard of Lyon and Adalbert of Prague, "people closed their eyes and everything was tolerated in exchange for good gold dinars."[355] This tolerance was encouraged by the elites of Western Europe, who had been using up the reserves of gold left over from Roman times. By the eighth century those reserves were all but gone, and it became difficult to buy luxury goods from the Middle East, particularly clothes, upholstery, tapestries, and carpets. The trade in young women and beardless boys provided scarce foreign exchange at an opportune time.[356]

The captives were taken overland by three routes: 1) through Germany and France to Muslim Spain; 2) through Venice and by ship to the Middle East; and 3) down the Dnieper, the Don, or the Volga to the Middle East via the Black Sea or the Caspian. How many were exported? Somewhere between several tens of thousands and several hundreds of thousands from the eighth century to the twelfth.[357] Figures are hard to come by because the trade was at the limits of legality and left few written records. Enslavement itself was legal, being seen as a humane way to dispose of criminals, prisoners of war, and others who would otherwise be put to death. In this case, however, the traffickers did more than enslave. They broke secular and Christian law by castrating boys, by breaking up families, and by abandoning the very old and the very young. Above all, they forbade their captives to join the Christian faith and exiled them to the Muslim world. For such captives, the existing term *servus* seemed inappropriate, and the ethnonym 'Slav' took its place to mean an especially degraded servant … a slave.[358]

As Western Europe's balance of trade became positive, its elites could spend more, and the old trading networks came back to life. Thus

---

[354] Skirda 2010, pp. 85-86.
[355] Skirda 2010, p. 75; see also Phillips 1985, pp. 59-64.
[356] Phillips 1985, p. 60; Skirda 2010, pp. 56-58.
[357] Skirda 2010, p. 6.
[358] Skirda 2010, pp. 94-100.

began a long economic recovery that would pull their societies out of the Dark Ages.

Decline and resurgence

The white slave trade declined during the Late Middle Ages. On the one hand, it became disreputable as Eastern Europe embraced Christianity. How could one enslave fellow believers for sale to Muslims? On the other hand, the Reconquista of Sicily, Spain, and several North African port cities eliminated the closest markets. The balance of power was shifting. European states were stronger and less willing to accept what had once been shameful but expedient.

Nonetheless, the end was long in coming. The trade declined from the twelfth to fifteenth centuries but then underwent a burst of renewed growth. The middlemen were no longer merchants buying unwanted prisoners of war but rather abductors raiding villages on the shores of the Mediterranean or the steppes of the Ukraine. All this was due to a Muslim resurgence under the Ottoman Turks. In the West, they pushed into North Africa and joined with local emirs to dislodge the Spaniards from Tunis, Algiers, and Oran. In the North, they advanced into the Balkans and allied themselves with the Crimean Tatars to seize the northern Black Sea coast. The western and northern frontiers became havens for slave-raiding states under Ottoman protection: the beyliks of North Africa and the Khanate of the Crimea. More than a million people would be abducted from Western Europe and between a million and a half and three million from Eastern Europe.[359]

In their new homeland, the enslaved women were so renowned for their beauty that the word 'white' came to mean 'beautiful' or 'valuable.' An eighth-century Arab musician wrote: "They used not to train beautiful slave girls to sing, but they used only to train yellow and black girls. The first to teach valuable girls to sing was my father."[360] The value of light female skin was reflected in slave prices: while women sold for the same price as men when the captives were black African, they sold for fifty

[359] Davis 2004; Fisher 1972; Kolodziejczyk 2006; Lewis 1990, pp. 11-13, 56, 79-80, 91; Skirda 2010, p. 6.
[360] Lewis 1990, p. 56.

percent more when the captives were Russian or Circassian.[361] A mid-nineteenth century report stated that a "trained, strong, black slave" cost 4,000 to 5,000 piasters in Turkey, whereas "white slave girls of special beauty" were worth 50,000 piasters or more.[362]

Many entered the highest strata of Muslim society, including Roxelana—the harem slave who became Suleiman the Magnificent's wife. Marriages were often arranged in this manner, as a Tunis notable explained when questioned in 1891:

> Q. Do you have any white slave women of Circassian or other origins?
> A. No, I don't have any.
>
> Q. When you went to Constantinople, you didn't bring any white slave women back with you?
> A. I brought two women back with me, that is true, but they aren't slaves. They are two sisters, one of whom is my wife. One is called Zohra and the other Daïde.
>
> Q. If they aren't slaves, how did you get them?
> A. I bought them in Constantinople, I freed them, and I married one of them. The other one is my sister-in-law.[363]

The marriages produced offspring with lighter skin—a visible mark of slave ancestry, and yet one worn with pride in the noblest of families. As an historian of Islam observed, "the Circassian women [...] have so curiously tinted the upper classes of oriental Muslim society."[364] The same observation was made by a seventeenth-century traveler about the Persian upper classes:

> [...] the Persian blood is now highly refined by frequent intermixtures with the Georgians and the Circassians, two nations which surpass all the world in personal beauty. There is hardly a man of rank in Persia who is not born of a Georgian or Circassian mother; and even the king himself is commonly sprung, on the female side, from one or other of these countries. As it is long since this mixture commenced, the Persian women

---

[361] Verlinden 1977, pp. 211, 224, 306, 315, 330-331, 460, 517; Frost 1990.
[362] Lewis 1990, p. 13.
[363] Dali 2012, pp. 349-350.
[364] Lévi-Provençal 1953, p. 179. See also Dali 2012.

> have become very handsome and beautiful, though they do not rival the ladies of Georgia.[365]

An Arab man could thus be eulogized as "the son of a white woman," and others would boast of having taken white women as captives.[366] Whiteness was the apex of femininity.

Endgame

The early seventeenth century was the peak of slave raiding in the western Mediterranean. From their bases in North Africa, the Barbary corsairs, as they were known, targeted islands and coasts to the north, sometimes striking as far away as Ireland and Iceland. Britain and France tried to rein them in by negotiating treaties with each beylik, but enforcement proved difficult. The bey received a share of the proceeds but could not easily restrain the corsairs, who saw him not as a leader to obey but as a backer to pay off. The raiding stopped when France conquered Algeria in 1830.[367]

Events played out similarly at the other end of Europe. In the early seventeenth century the Don Cossacks began launching attacks on the Khanate of the Crimea.[368] This local conflict eventually merged into the Great Turkish War (1683-1699), when Austria, Poland, Russia, and Venice joined forces to defeat the Ottoman Empire and its Tatar allies. The peace treaty included a provision to end the slave trade, but enforcement, as always, proved difficult because the alternatives of farming or herding held less appeal. The trade finally ended in the Crimea when Russia annexed the region in 1783.[369] For a few more decades, it continued on a small scale in the Caucasus, where many women willingly offered themselves for sale in the hope of marrying a wealthy Turk or even the sultan.[370]

With fewer women being taken from Europe, more were brought to market from sub-Saharan Africa. That traffic had hitherto been largely

---

365 Lawrence 1848, p. 310.
366 Lewis 1990, p. 113.
367 Davis 2004.
368 Fisher 1972; Fisher 1973; Kisilov 2007.
369 Abou-el-Haj 1969; Kisilov 2007.
370 Millingen 1870-1871.

male: men had outnumbered women two to one among slaves crossing the Atlantic to the Americas[371] and predominated among those crossing the Sahara to North Africa.[372] The trans-Saharan traffic now became overwhelmingly female.[373] Especially popular were Fulani, Nubian, and Abyssinian women, who looked less African in their skin color, hair length, and facial features.[374]

Figure 15: *Cairo slave market*, by Maurycy Gottlieb (1856-1879)

Has the same sort of trade existed elsewhere?

There have been other cases of women being marketed transnationally for their fair skin, although none of them rivals the medieval and post-medieval trade in European women. One case was nineteenth and early twentieth-century Thailand, where wealthy Chinese often gave up their daughters in exchange for protection. Such women were "prized for their light skin color." In Thailand, "skin colour was valued as a form of feminine beauty and a sign of 'Chineseness.'"[375]

Meanwhile, East Asia and Southeast Asia were seeing a rise in transnational prostitution. That trade involved lighter-skinned women, specifically Chinese, Japanese, Eurasians, and Russians from east of the

---

371 Manning 1983.
372 Bresc 2012.
373 Lewis 1990, pp. 12, 91, 72-77; Manning 1983.
374 Lewis 1990, pp. 74-75; Renault and Daget 1985, p. 59.
375 Jiemin 2003.

Urals.[376] In addition to serving the East Asian market, Chinese and Japanese prostitutes also went south to Indochina, the Malay states, Singapore, and the Dutch East Indies.[377] In recent times Chinese sex trafficking has again spread to Southeast Asia, particularly Malaysia:

> Over the years, she [a Malaysian Indian sex worker] has seen how Malaysian men, who generally could not afford to pay Chinese migrant women's rates, would save up just to experience a foreign woman with a lighter skin tone.
>
> [...] Southeast Asian women were the preferred nationalities for Syndicate X, its counterparts, and clientele during the early 2000s in KL [Kuala Lumpur]. Since then, they have been displaced by women from the PRC [People's Republic of China] and women who look "white."[378]

Chinese sex trafficking is also spreading to sub-Saharan Africa.[379] In Douala, Cameroon, "the cheap and affordable commoditised sex proposed by these young Chinese women enables relatively poor local men to partake in the global economy of body consumption."[380]

To sum up, few East Asian women were traded transnationally before the nineteenth century. One reason was the region's relative isolation within the world trading system. Another was the lack of nearby markets where light-skinned women would be not only scarce but also sought by large numbers of men with disposable income. All of that has changed with the growing affluence of Southeast Asia and sub-Saharan Africa, as well as the integration of China into the global economy.

---

376 League of Nations 1934.
377 Kligman and Limoncelli 2005.
378 Chin 2013, pp. 91-92, 106, 178.
379 Ndjio 2009; Ndjio 2014a. Ndjio has claimed that between 13,000 and 18,500 Chinese prostitutes are plying their trade in sub-Saharan Africa, but that claim is considered to be exaggerated. See French.CHINA.ORG.CN 2016.
380 Ndjio 2014b.

Head hair: longer, silkier, and more diverse forms

We now return to head hair. It became not only more brightly colored among Europeans but also longer and silkier, again through an evolutionary change that began with women. In this case, however, the change apparently took place throughout the whole of northern Eurasia, and not just in Europe.

A competitive mate market will reward features that are not only brighter but also bigger, longer, smoother, or flashier. This kind of supernormal stimulus has been studied in many species. For instance, a male butterfly is attracted to the flashing wing pattern of a passing female, even if the pattern is printed on a rotating drum, and its response will increase as the speed of rotation increases—up to almost ten times the speed of a normal wing-beat.[381] Why does the male bother responding to stimuli that do not exist in nature? The answer should be obvious. There is no need to program an upper limit into the male's response if the stimulus normally has an upper limit.

In the case of our species, head hair has become much longer than hair elsewhere on the body, as if it were a supernormal stimulus that holds some power of attraction over the observer. Charles Darwin argued that long head hair "probably serves as an ornament and was acquired through sexual selection."[382] The lengthening was due to several evolutionary changes: faster rate of growth, longer growing phase, higher density, and greater resistance to physical damage.[383] Such changes went farther in some populations than in others. Darwin noted "the extraordinary difference in the length of the hair in the different races; in the negro the hair forms a mere curly mat; with us it is of great length, and with the

[381] Manning 1972, pp. 47-49. The species is the silver-washed fritillary butterfly.
[382] Darwin 1936 [1888], p. 906.
[383] Franbourg et al. 2003; Khumalo 2005; Loussouarn 2001; Loussouarn et al. 2005.

American natives it not rarely reaches to the ground."[384] He concluded that "long tresses" must have been "a late acquisition" in human evolution.[385]

Short, frizzy hair is the ancestral form. Today, it is seen in sub-Saharan Africans and in some remnant groups that remain in the tropical regions of South Asia, Southeast Asia, and parts of Oceania, notably the Andamanese of India, the Semang of Malaysia, the Aeta of the Philippines, and the natives of Australia, Papua New Guinea, and Melanesia.

Long hair is the derived form. It evolved in those humans who moved into northern Eurasia, including some who later moved back to the Tropics, such as Austronesians in Southeast Asia and Oceania and Amerindians in the tropical New World.

Long head hair, a component of the *Kindchenschema*

The ancestral form is straight and silky only in newborn children: "[...] the majority of African babies are not born with springy tight curls, the African child at birth is either bald or has silky loose curls similar to the Jheri curls."[386] A survey of Zambian students found that they viewed long hair and white skin as two features that African babies have in common with Europeans.[387] Loose, silky hair seems to have been initially specific to infants and, as such, formed part of the *Kindchenschema*—the visual, auditory, and tactile features that identify a human infant and which are perceived as "cute."

That perception is universal among humans. Yet, no matter how strong it may be, it is not sufficient to create naturally long hair in a population. There must also be a competitive mate market to translate that perception into sexual selection. Indeed, head hair is naturally short in some populations where women nonetheless make strenuous efforts to lengthen it, either by applying heat or by supplementing with fibers and grasses. These practices are attested by elderly African women who

---

384 Darwin 1936 [1888], p. 906.
385 Darwin 1936 [1888], p. 906.
386 Ajose 2012.
387 Powdermaker 1956.

recount how they used to straighten their hair with hot stones before hot combs became available. There is also the testimony of comb-like artefacts from archaeological sites. [388]

Such practices were noted by early European observers. In 1721, John Atkins described women braiding and dressing their hair in Sierra Leone:

> [The women] work hard at Tillage, make Palm-Oil or spin Cotton, and when they are free from such work, the idle Husbands put them upon braiding, and fettishing out their woolly hair, (in which Sort of Ornament they are prodigious proud and curious) keeping them every Day, for many Hours together at it.[389]

West African women still lavish much time and effort on their hair:

> "Big hair," "plenty of hair," "much hair"—West African communities, including Mende, admire a fine head of long, thick hair on a woman. Both these elements are crucial: thickness and length. Thickness equals increase in the number of individual strands, and the length is proof of strength. Growing such luxuriant hair requires a Mende woman's patience and care. Because a man's hair is kept shaved or cut close to the scalp, people say that "men don't have hair." Beautiful hair thus is a distinctly female trait; the more of it, the more feminine the woman.[390]

Hairdressing is old enough in West Africa to have spawned myths, such as this one:

> It is known among Mende that all the "water people," angels, have marvelous hair. The mermaid *Tingoi* is known by her long, wavy hair and her glamorous habit of dressing it with a golden comb while seated on a rock. A little girl with especially long hair is feared to be in danger of drowning because she will be very attractive to the "water people," who may think she is one of them and wish her to join them.[391]

---

[388] Khumalo 2008. Also see Sieber and Herreman 2000.
[389] Sieber and Herreman 2000, p. 67.
[390] Boone 1986, p. 184.
[391] Boone 1986, p. 192.

Long-haired women appear in the folklore of other African peoples. Among the Yoruba, a folk-tale explains "why women have long hair." A woman fell into a pit and was pulled out by her hair, which thereby became as long as a man's arm. She initially felt ashamed of her new appearance and hid herself.

> But after a while she realized that her long hair was beautiful, and then she felt very proud and scorned all the short-haired women, jeering at them. When they saw this, they were consumed with jealousy, and began to be ashamed of their short hair. "We have men's hair," they said to one another. "How beautiful it would be to have long hair!"
>
> So one by one they jumped into the pit, and their friends pulled them out by the hair.
>
> And in this way they, and all women after them, had long hair.[392]

Another Yoruba folk-tale recounts how a king had a beautiful daughter with hair "so long that it touched the ground when she walked." But she lost her hair and regained it only after a man found a tree that bore human hair. She then became his wife.[393]

During the early colonial era, ethnographers observed lengthening of women's hair throughout West Africa and to a lesser extent in central and southern Africa. Women of the Manyema (Tanzania) were described as having "an abundance of hair" that would "flow down to the waist in masses of ringlets."[394] Among the Sango (DRC), girls old enough for marriage would plait long strands of string into their hair to create long manes.[395] Young women of the Mbalantu (Namibia) achieved the same effect by braiding sinew extensions.[396] Such practices were apparently uncommon in South Africa. When a black South African woman traveled to the United States in the 1930s, she was struck by the number of African American women who straightened and extended their hair. "What really

---

[392] Ogumefu 1929, chap. V.
[393] Abrahams 1983, pp. 59-63.
[394] Bettany 1892, p. 661.
[395] Friedrich 1913, pp. 190-191.
[396] Sieber and Herreman 2000, p. 65.

made me feel strange [was] nearly every girl and woman has long hair and I among them looked like a boy dressed in girl's clothes."[397]

African American women were braiding and threading their hair at an early date, a practice they may have brought with them from West Africa. Some would straighten their hair by first greasing it and then applying a heated cloth. In contrast, men often shaved their heads, an indication that long head hair was viewed as a female ornament.[398]

In sum, African women lengthened their head hair to look feminine, and not "white." The motive was not self-hatred but sexual fantasy—a desire for the supernormal, a wish to become a woman with a long mane of hair. Humans can and will create by artificial means what has been created elsewhere through biological evolution. The same desire is satisfied differently.

Figure 16: Long-haired Sango woman, Democratic Republic of the Congo (Friedrich 1913, Fig. 174).

An evolutionary trajectory

African hair can be reverted to its loose, silky form of infancy by some diseases: AIDS, rheumatoid arthritis, systemic lupus erythematosus, pulmonary tuberculosis with cachexia, and Behçet's disease.[399] Normal hair growth is somehow disrupted. A similar disruption of genetic origin, that is, a loss-of-function allele, may have set the stage for the evolution of long head hair.

Such a mutation then had to spread through the population at

[397] Thomas 2006, p. 487.
[398] White and White 1995.
[399] Ajose 2012.

the expense of short hair. This process apparently began outside Africa, in non-tropical environments that shifted the pressure of sexual selection from men to women.[400] Women were now in the spotlight, and their visible features became larger or flashier. In the case of head hair, the way forward might have been shown by the practice of hair lengthening: once that practice was in place, selection began to favor women whose hair could be lengthened with less effort. Over succeeding generations the infant hair form persisted more and more into adulthood while becoming longer and straighter, eventually reaching the waist if left uncut. The cultural became biological.

That lengthening happened not long after modern humans entered northern Eurasia some 30,000 years ago—the dating of the *EDAR* allele that produces thick straight hair in East Asians.[401] In Europeans, straight hair is produced by alleles at other genes, notably *TCHH*, *WNT10A*, and *FRAS1*. It was initially thought that straight hair evolved independently in Europeans and East Asians.[402] Ancient DNA now suggests a somewhat different scenario:

1. In a population ancestral to Europeans and East Asians, hair became thick, straight, and long some 30,000 years ago via a derived allele at the *EDAR* gene.

2. Those ancestral Eurasians split into two groups: a western group that would become present-day Europeans and an eastern group that would become present-day East Asians.[403]

3. Thick straight hair remained prevalent in the eastern group but disappeared in the western group. Nonetheless, it was still present in half of Europeans as late as 8,000 years ago, according to ancient DNA from Motala, Sweden.[404] Today, its incidence is 87% in East Asians and about 1% in Europeans.[405]

[400] See second chapter of this book.
[401] Kamberov et al. 2013.
[402] Medland et al. 2009; Pospiech et al. 2015; Tan et al. 2013.
[403] Rogers 1986.
[404] Mathieson et al. 2015.
[405] Unterländer et al. 2017.

4. Europeans thus went from thick straight hair to thin hair with diverse forms: ~45% now have straight hair, ~40% wavy hair, and ~15% curly hair.[406] Curly hair is thus a derived trait among Europeans and not a holdover from ancestral Africans. Hair form became diverse in Europeans probably for the same reason that hair and eye color did—a desire for novelty. Indeed, an Austrian study has shown that women tend to change their hair form to a less common one.[407]

In sum, the evolutionary trajectory may have begun with women pushing the bounds of their phenotype by artificially lengthening their head hair. That artificial phenotype provided sexual selection with a template when humans moved beyond the tropical zone and into environments where they experienced stronger sexual selection. Those who found mates had less trouble lengthening their hair because it had naturally become less brittle and potentially longer.

Hair became longer over the whole of northern Eurasia. Only within the "epicenter" of sexual selection, in northeastern Europe and western Siberia, was the pressure of selection so intense that it not only lengthened hair but also diversified its forms and colors.

Although women were the ones being selected for longer hair, sexual selection also lengthened the hair of men because the genes in question are weakly sex-linked, as are most genes. Nonetheless, there is some sex linkage. Scalp hairs have a greater mean diameter and hence more volume in women, even in shorter-haired New Guineans.[408] Hair growth rate and final length are also somewhat greater in women than in men.[409] Men furthermore tend to lose their head hair, often as early as their twenties. In general, growth of head hair is under stronger hormonal inhibition in men than in women.[410]

As Charles Darwin concluded, head hair serves an ornamental purpose: "for we know that long tresses are now and were formerly much

---

406 Medland et al. 2009.
407 Schweder 1994.
408 Walsh and Chapman 1966.
409 Valković 1988, p. 6, cited in Sigler 2011, p. 13.
410 Kondo et al. 1990.

admired, as may be observed in the works of almost every poet; St. Paul says, 'if a woman have long hair, it is a glory to her.'"[411]

[411] Darwin 1936[1888], p. 906.

Face shape, eye morphology, waist-to-hip ratio ...

Finally, we come to several other visible traits: face shape, eye morphology, shoulder width, waist-to-hip ratio ... The pattern is similar. Europeans diverged from other populations through a process of selection that targeted certain visible features of women, particularly on or near the face.

Face shape

A comparative study of European and West African face shape ascribed the differences between the two groups to a selection pressure that acted primarily on women.[412] As with the other traits we have discussed, selection for more feminine-looking women may have spilled over to the other sex, thus feminizing the faces of the whole population.

Facial feminization is linked to European eye colors. This was an unintended finding of two Czech studies whose participants were asked to rate male and female facial photos. Initially, the brown-eyed male faces were rated as more dominant than the blue-eyed male faces. When, as a control, the brown-eyed faces were photoshopped to make them blue-eyed, they were still rated as more dominant. On careful examination, the originally brown-eyed faces were found to be more masculine with broader and more massive chins, broader mouths, larger noses, larger eyebrows, and closer-set eyes. The originally blue-eyed faces had smaller and sharper chins, narrower mouths, smaller noses, and greater distance between the eyes. Blue eyes were associated with a more feminine face shape only if the participants were male. If they were female, eye color was not associated with facial feminization, perhaps because a female fetus normally has enough estrogen in its body tissues to feminize its face,

---

[412] Liberton et al. 2009.

whereas a male fetus normally does not. If a male fetus develops blue eyes, it more likely has a high enough level of estrogen to trigger facial feminization.[413]

Did eye color correlate with face shape in the above Czech studies because some of the brown-eyed men were partly Jewish or Roma and had a more Mediterranean appearance? If that were the case, face shape would have been more variable in the brown-eyed men, yet it was not. This explanation also fails to explain the influence of sex: why did blue eyes correlate with facial feminization in men but not in women?

On a perhaps related note, a correlation has been observed between blue eyes and shyness. A study of preschoolers found more social wariness in blue-eyed boys than in brown-eyed boys. The difference was greatest at the extremes of wariness. Among the very inhibited boys, 13 out of 14 were blue-eyed. Among the very uninhibited, only 4 out of 10 were. As with the correlation between blue eyes and facial feminization, there was no such relationship among the girls, whose eyes were blue in 5 out of 9 among the very inhibited and in 6 out of 11 among the very uninhibited. The proportions were close to parity in both cases.[414]

Another study confirmed the male-only relationship in children from kindergarten and the second grade but not in those from the fourth grade.[415] A study of adults had a similar finding, although the distinction was between participants with light eyes (hazel, green, grey, blue) and those with dark eyes (brown, black).[416]

Eye morphology

The eyes of Europeans look rounder in women than in men. This sex difference exists because the exposed sclera—the white of the eye—is more horizontally exposed in men. In addition, eye fissures are narrower and less rectangular in women than in men. These sex differences are absent in other human populations.[417]

---

[413] Kleisner et al. 2010; Kleisner et al. 2013. See also Figure 17.
[414] Coplan et al. 1998; See also Rosenberg and Kagan 1987.
[415] Rubin and Both 1989.
[416] Bassett and Snyder 2010.
[417] Danel et al. 2018; Danel et al. 2020.

The authors of these findings considered the explanation of sexual selection but rejected it because female eye morphology does not correlate with two other aspects of female attractiveness: face shape and facial averageness.[418] The lack of correlation, however, may simply mean that these three aspects impose different constraints on the direction of sexual selection:

Eye morphology - sexual selection is unidirectional. Women are more attractive if they have rounder eyes.

Face shape - the direction of sexual selection goes into reverse beyond a certain point. Women are more attractive if they have smaller chins and smaller noses, but only up to a certain point. Excessively small chins and noses are not attractive either.

Facial averageness - the constraints are again different. Women become less attractive on each side of a narrow median.

Shoulder width and waist-to-hip ratio

A Latvian study has found small but significant correlations between female eye color and certain sexually dimorphic features. Shoulder width and waist-to-hip ratio had lower values among blue-eyed women than among brown-eyed women.[419] Indeed, several studies have reported that women of European descent have narrower waists, wider hips, and thicker subcutaneous fat than do women of other origins.[420]

Future research

To explain the higher incidence of certain diseases among red-haired women, it has been argued that the development of hair and eye colors is determined not only by specific alleles but also by the degree to which the hair follicles and the iris are exposed to estrogen during fetal development. The higher the level of estrogen, the brighter the child's hair and eye color, with red hair being the most estrogen-dependent.[421] The role

---

418 Danel et al. 2020.
419 Kažoka and Vētra 2011.
420 Choi and Trotter 1970; Gasperino 1996; Hrdlicka 1898; Meredith and Spurgeon 1980; Nelson and Nelson 1986.
421 Frost, Kleisner, and Flegr 2017.

of genetics still matters: some combinations of alleles are more likely to produce blue eyes, and others more likely to produce brown eyes.

If eye color is sex-linked via fetal exposure to the sex hormones, it should correlate with the digit ratio—the length of the index finger divided by the length of the ring finger. The digit ratio roughly measures the ratio of estrogens to androgens in the womb: the lower the value, the more the fetus was masculinized; the higher the value, the more it was feminized.[422] An unpublished study, using a sample of 644 participants, found a significant effect of left-hand digit ratio on hair color and, to a lesser extent, eye color. Specifically, blond hair was associated with a higher mean digit ratio than were brown hair, red hair, and other hair colors. Left-hand digit ratio had a significant but weaker effect on eye color: blue eyes were associated with a higher mean digit ratio than were brown eyes and other eye colors.

That study was not published because of two objections from the referees: hair dyeing could not be excluded as a possible factor; and identification of hair and eye color was too subjective. Yet it is difficult to see how hair dyeing or misidentification of hair and eye color can explain the reported associations between digit ratio and hair or eye color. If anything, such methodological problems would introduce more noise into the data and make the associations less significant.

Five points for future research:

1. The sample size should be large.
2. The stronger effect of left-hand digit ratio suggests that the sex hormones can affect hair and eye color both prenatally and postnatally. Right-hand digit ratio is a better measure of prenatal influences taken alone.
3. Because hair and eye color darken with age, the sample should have a narrow age range.
4. Ethnic differences may have to be controlled.
5. The sex hormones have prenatal effects on hair color that are the opposite of their pubertal effects on hair color. Whereas women are lighter-haired than men from 17 onward, they are

[422] See, for example, Manning et al. 2004.

actually darker-haired up to the age of 14.[423] Consequently, right-hand digit ratio should be better at predicting the darkening of hair color before puberty, and left-hand digit ratio better at predicting the lightening of hair color after puberty. The age range should therefore affect the results of any study on the relationship between the sex hormones and hair color.

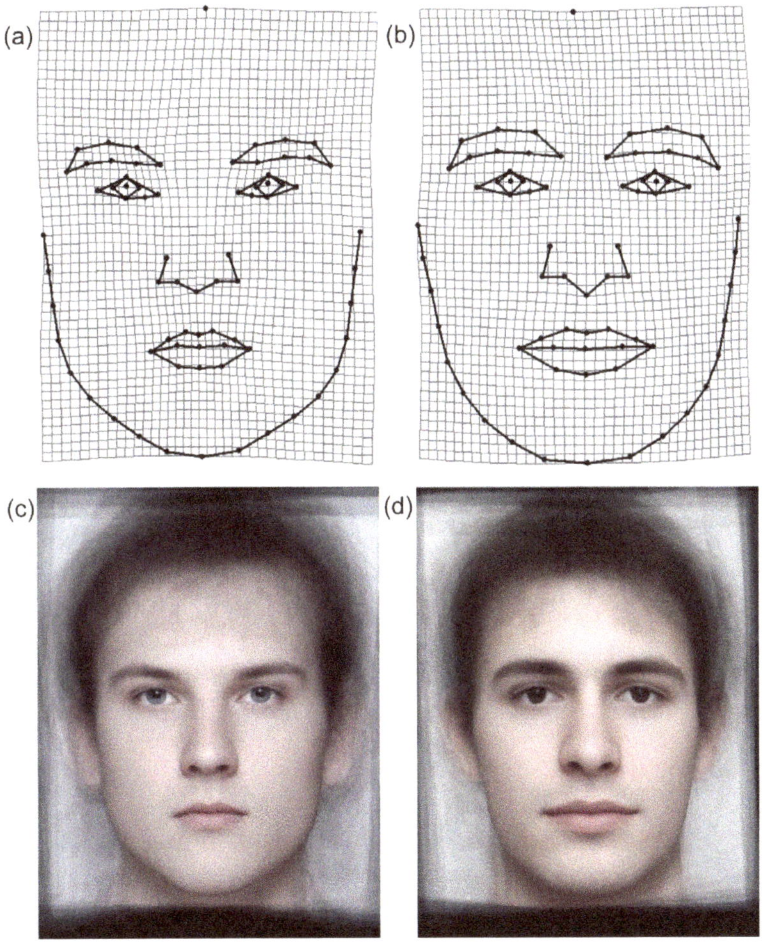

Figure 17: Averaged faces of blue-eyed (left) and brown-eyed (right) male participants. Czech population (Kleisner et al. 2010). Even among people of the same ethnic background, blue eyes tend to coexist with a different facial structure. The face looks feminized.

423 Steggerda 1941.

Parting thoughts

Our species has diversified considerably in its outward appearance, with some of the most striking changes being relatively recent—during the time of the last ice age, on the plains stretching from northeastern Europe to western Siberia. That time and space saw a new color scheme arise on the human body: the hair and eyes independently took on bright hues and the skin turned white. The changes began on the visible surfaces of the female body, particularly the face—the focus of visual attention. This visual transformation then spread to the male body, as the new color traits were weakly sex-linked and had little pushback from natural selection.

Those changes happened because men were scarce, their ranks depleted by the hazards of hunting over long distances, and because only a few of the ablest hunters could afford to take a second wife. Conversely, women were plentiful on the mate market. Men were able to pick and choose, and they chose the most eye-catching women.

There thus appeared a radically new look. It would twice come close to extinction: at the height of the last ice age, when human life dwindled away on the frozen wastes, and later when a warming climate caused the steppe-tundra and the reindeer herds to disappear. Yet it survived, eventually spreading to the entire continent and becoming identified with Europe, all this on the eve of recorded history ... as if, moments before curtain time, a cast of actors had been hastily made up and rushed onto the stage.

References

Abbie, A.A. and W.R. Adey. (1953). Pigmentation in a central Australian tribe with special reference to fair-headedness. American Journal of Physical Anthropology 11(3), 339-359.
https://doi.org/10.1002/ajpa.1330110310

Abou-el-Haj, R.A. (1969). The Formal Closure of the Ottoman Frontier in Europe: 1699-1703. Journal of the American Oriental Society 89(3): 467-475.
https://doi.org/10.2307/596616

Abrahams, R.D. (1983). African Folktales. Traditional Stories of the Black World. New York: Pantheon Books.

Ainger, S.A., K. Jagirdar, K.J. Lee, H.P. Soyer, and R.A. Sturm. (2017). Skin pigmentation genetics for the clinic. Dermatology 233: 1-15.
https://doi.org/10.1159/000468538

Ajose, F.O.A. (2012). Diseases that turn African hair silky. International Journal of Dermatology 51 (supp. S1): 12-16.
https://doi.org/10.1111/j.1365-4632.2012.05556.x

Akande, H. (2015). Illuminating the Difference. Black, White, and Brown Women. London: Rabaah Publishers.

Albert, E.M. (1963). Women of Burundi. A study of social values. In D. Paulme (Ed.) Women of Tropical Africa (pp. 179-216). London: Routledge & Kegan Paul.

Ali, M.M. and V. Vaidya. (2007). Vitamin D and cancer. Journal of Cancer Research and Therapeutics 3(4): 225-230
https://doi.org/10.4103/0973-1482.38998

Alley, T. R. (1980). Infantile colouration as an elicitor of caretaking behaviour in Old World primates. Primates 21(3): 416-429.
https://doi.org/10.1007/BF02390470

Alley, T.R. (2014[1986]). An ecological analysis of the protection of primate infants. In V. McCabe and G.J. Balzano (Eds.) Event Cognition: An Ecological Perspective. Routledge.

Almedingen, E.M. (1957). Russian Fairy Tales. London: Frederick Muller Ltd.

Anderson, W.W. (1969). Polymorphism resulting from the mating advantage of rare male genotypes. Proceedings of the National Academy of Sciences USA 64(1): 190-197.
https://doi.org/10.1073/pnas.64.1.190

Andresen, T., D. Lunden, A.M. Drewes, and L. Arendt-Nielsen. (2011). Pain sensitivity and experimentally induced sensitisation in red haired females. Scandinavian Journal of Pain 2(1):3-6.
https://doi.org/10.1016/j.sjpain.2010.08.005

Angel, J.L. (1972). Biological relations of Egyptian and eastern Mediterranean populations during Pre-dynastic and Dynastic times. Journal of Human Evolution 1(3): 307-313.
https://doi.org/10.1016/0047-2484(72)90066-8

Anon. (2008a). Bygone brunette beauty: Fashion in hair color. Gene Expression, June 29.
http://www.gnxp.com/blog/2008/06/bygone-brunette-beauty-fashion-in-hair.php

Anon. (2008b). Maxim's audience prefers brunettes; distribution is bimodal. Gene Expression, July 6.
http://www.gnxp.com/blog/2008/07/maxims-audience-prefers-brunettes.php

Anon. (2015). The Archeological Museum of Thessaloniki. Report on ancient DNA -Learn what eye color your ancestor had and what he ate in the Neolithic! Iefimerida
http://www.iefimerida.gr/news/219751/ekthesi-gia-arhaio-dna-mathe-ti-hroma-matia-eihe-kai-ti-etroge-o-neolithikos-progonos

Aoki, K. (2002). Sexual selection as a cause of human skin colour variation: Darwin's hypothesis revisited. Annals of Human Biology 29(6): 589-608.
https://doi.org/10.1080/03014460210000019144

Ardener, E.W. (1954). Some Ibo attitudes to skin pigmentation. Man 54: 71-73.
http://www.jstor.org/stable/2793760

Asher, S.R. and V.L. Allen. (1969). Racial preference and social comparison processes. Journal of Social Issues. 25(1): 157-166.
https://doi.org/10.1111/j.1540-4560.1969.tb02584.x

Ashikari, M. (2005). Cultivating Japanese whiteness. The 'whitening' cosmetics boom and the Japanese identity. Journal of Material Culture 10(1):73-91. https://doi.org/10.1177/1359183505050095

Atzenhoffer, R. (2011). Les hommes préfèrent les blondes. Les lectrices aussi. Effet de psychologie, horizons idéologiques et valeurs morales des héroïnes dans l'œuvre romanesque de H. Courths-Mahler. Colloque national (CNRIUT), Villeneuve d'Ascq, 8-10 juin 2009, http://cnriut09.univ-lille1.fr/articles/Articles/Fulltext/10a.pdf

Babcock, L. (1929). Tan takes its turn as a maker of fashion. The Delineator, February, p. 25.

Bae, B-I., D. Jayaraman, and C.A. Walsh. (2015). Genetic changes shaping the human brain. Developmental Cell 32: 423-434. https://doi.org/10.1016/j.devcel.2015.01.035

Baird, D.T. (1976). Oestrogens in clinical practice. In J.A. Loraine and E. Trevor Bell (Eds.) Hormone assays and their clinical application, (p. 408). Edinburgh: Churchill Livingstone.

Balikci, A. (1967). Female infanticide on the Arctic coast. Man 2(4): 615-625. https://doi.org/10.2307/2799344

Balloux F., L.J. Handley, T. Jombart, H. Liu, and A. Manica. (2009). Climate shaped the worldwide distribution of human mitochondrial DNA sequence variation. Proceedings of the Royal Society B. Biological Sciences 276: 3447-3455.
https://doi.org/10.1098/rspb.2009.0752

Bancroft, I.R. (1944). Hormones and the skin. California and Western Medicine 61(2): 60-62.

Bannatyne, R.M., P.N. Skowron, and J.L. Weber. (1969). Job's syndrome-a variant of chronic granulomatous disease: Report of a case. The Journal of Pediatrics 75(2): 236-242.
https://doi.org/10.1016/S0022-3476(69)80394-X

Bard, C. (1998). Les garçonnes. Modes et fantasmes des Années folles. Paris: Flammarion.
https://doi.org/10.4000/clio.270

Barnes, J.A. (1951). Marriage in a Changing Society. Cape Town: Oxford University Press.

Baroja, J.C. (1964). The World of Witches. trans. by N. Glendinning, Weidenfeld and Nicolson.

Bartok, B. (1967). Rumanian Folk Music. Vol. 3. The Hague: Martinus Nijhoff.

Bartok, B. and A.B. Lord. (1951). Serbo-Croatian Folk Songs. New York: Columbia University Press.

Bassett, J.F., and T.L. Snyder. (2010). Eye color, behavioral inhibition, and performance in self-paced and reactive hunting domains: A further test of the Worthy hypothesis. Individual Differences Research 8(4): 239-245.

Baudouin, J.-Y., and R. Brochard. (2011). Gender-based prototype formation in face recognition. Journal of Experimental Psychology: Learning, Memory, and Cognition 37(4): 888-898.
http://dx.doi.org/10.1037/a0022963

Beleza, S., A.M. Santos, B. McEvoy, I. Alves, C. Martinho, E. Cameron, et al. (2013). The timing of pigmentation lightening in Europeans. Molecular Biology and Evolution 30(1): 24-35.
https://doi.org/10.1093/molbev/mss207

Bernstein, I.H., T-D. Lin, and P. McClellan. (1982). Cross- vs. within-racial judgments of attractiveness. Perception & Psychophysics 32 (6): 495-503. https://doi.org/10.3758/BF03204202.

Berry, S. (2000). Screen Style. Fashion and Femininity in 1930s Hollywood. Minneapolis: University of Minnesota Press.

Best, D.L., J.T. Field, and J.E. Williams. (1976). Color bias in a sample of young German children. Psychological Reports 38(3): 1145-1146.
https://doi.org/10.2466/pr0.1976.38.3c.1145

Best, D.L., C.E. Naylor, and J.E. Williams. (1975). Extension of color bias research to young French and Italian children. Journal of Cross-Cultural Psychology 6(4): 390-405.
https://doi.org/10.1177/002202217564003

Bestelmeyer, P.E.G., B.C. Jones, L.M. DeBruine, A.C. Little, D.I. Perrett, A. Schneider, L.L.M. Welling, C.A. Conway. (2008). Sex-contingent face aftereffects depend on perceptual category rather than structural encoding. Cognition 107(1): 353-365.
https://doi.org/10.1016/j.cognition.2007.07.018

Béteille, A. (1967). Race and descent as social categories in India. Daedalus 96(2): 444-463.
http://www.jstor.org/stable/20027046

Bettany, G.T. (1892). The World's Inhabitants or Mankind, Animals, and Plants. London: Ward, Lock, Bowden and Co.

Binford, S.R. (1968). A structural comparison of disposal of the dead in the Mousterian and the Upper Paleolithic. Southwestern Journal of Anthropology 24(2): 139-151
https://doi.org/10.1086/soutjanth.24.2.3629419

Birket-Smith, K. (1929). The Caribou Eskimos. Material and social life and their cultural position. Report of the 5th Thule Expedition 1921-24, Vol. 5. Copenhagen: Nordisk Forlag.

Blaffer-Hrdy, S. (2000). Mother Nature: Maternal Instincts and How They Shape the Human Species. New York: Ballantine.

Blaffer-Hrdy, S. and J. Hartung. (1979). The evolution of sexual dichromatism among primates. American Journal of Physical Anthropology 50(3): 450.

Blümner, H. (1892), Die Farbenbezeichnungen bei den Romischen Dichtern, Berlin: Stud. f. class.
https://doi.org/10.1524/phil.1889.48.14.706

Boone, S.A. (1986). Radiance from the Waters: Ideals of Feminine Beauty in Mende Art. New Haven and London.
https://doi.org/10.2307/3336598

Booth, C. (1962). Some observations on behavior of Cercopithecus monkeys. Annals of the New York Academy of Sciences 102(2): 477-487.
https://doi.org/10.1111/j.1749-6632.1962.tb13654.x

Boscoe, F.P., and M.J. Schymura. (2006). Solar ultraviolet-B exposure and cancer incidence and mortality in the United States. BMC Cancer 6: 264-314.
https://doi.org/10.1186/1471-2407-6-264

Bouakaze, C., C. Keyser, E. Crubézy, D. Montagnon, and B. Ludes. (2009). Pigment phenotype and biogeographical ancestry from ancient skeletal remains: inferences from multiplexed autosomal SNP analysis. International Journal of Legal Medicine 123(4): 315-325.
https://doi.org/10.1007/s00414-009-0348-5

Boule, M. and H.V. Vallois. (1957). Fossil Men. New York: Dryden Press.

Box, N.F., J.R. Wyeth, L.E. O'Gorman, N.G. Martin, and R.A. Sturm. (1997). Characterization of melanocyte stimulating hormone receptor variant alleles in twins with red hair. Human Molecular Genetics 6(11): 1891-1897. https://doi.org/10.1093/hmg/6.11.1891

Brace, C.L., N. Seguchi, C.B. Quintyn, S.C. Fox, A.R. Nelson, S.K. Manolis, and P. Qifeng. (2006). The questionable contribution of the Neolithic and the Bronze Age to European craniofacial form. Proceedings of the National Academy of Sciences U.S.A. 103(1): 242-247 https://doi.org/10.1073/pnas.0509801102

Brace, S., Y. Diekmann, T.J. Booth, Z. Faltyskova, N. Rohland, S. Mallick, et al. (2019). Ancient genomes indicate population replacement in Early Neolithic Britain. Nature Ecology & Evolution 3(5): 765-771. https://doi.org/10.1038/s41559-019-0871-9

Bramanti, B., M.G. Thomas, W. Haak, M. Unterlaender, P. Jores, K. Tambets, I. Antanaitis-Jacobs, M.N. Haidle, R. Jankauskas, C.J. Kind, et al. (2009). Genetic discontinuity between local hunter-gatherers and Central Europe's first farmers. Science 326: 137-140. https://doi.org/10.1126/science.1176869

Branicki, W., U. Brudnik, and A. Wojas-Pelc. (2009). Interactions between HERC2, OCA2 and MC1R may influence human pigmentation phenotype. Annals of Human Genetics 73(2): 160-170. https://doi.org/10.1111/j.1469-1809.2009.00504.x

Branicki, W., F. Liu, K. van Duijn, J. Draus-Barini, E. Pošpiech, S. Walsh, et al. (2011). Model-based prediction of human hair color using DNA variants. Human Genetics 129(4): 443-454. https://doi.org/10.1007/s00439-010-0939-8

Bresc, H. (2012). Esclaves noirs et esclaves blancs en Sicile (vers 1300-vers 1450). Entre déshumanisation et socialisation. In R. Botte and A. Stella (Eds.) Couleurs de l'esclavage sur les deux rives de la Méditerranée (Moyen Âge-XXe siècle), (pp. 55-84). Paris: Éditions Karthala. https://doi.org/10.4000/anneemaghreb.1106

Briot, E. (2007). Couleurs de peau, odeurs de peau : le parfum de la femme et ses typologies au xixe siècle. Corps 2(3): 57-63. https://doi.org/10.3917/corp.003.0057

Brockmole, J.R., and W.R. Boot. (2009). Should I stay or should I go? Attentional disengagement from visually unique and unexpected items at fixation. Journal of Experimental Psychology: Human Perception 35(3): 808-815. http://dx.doi.org/10.1037/a0013707

Brooks, R. (2002).Variation in female mate choice within guppy populations: population divergence, multiple ornaments and the maintenance of polymorphism. Genetica 116(2-3): 343-358.
http://dx/doi.org/10.1023/A:1021228308636

Bruce, V., and S. Langton. (1994). The use of pigmentation and shading information in recognising the sex and identities of faces. Perception 23(7): 803-822.
http://dx.doi.org/10.1068/p230803

Brues, A.M. (1975). Rethinking human pigmentation. American Journal of Physical Anthropology 43(3): 387-391.
https://doi.org/10.1002/ajpa.1330430320

Burch, Jr., E.S. (1972). The caribou/wild reindeer as a human resource. American Antiquity 37(3): 339-368.
https://doi.org/10.2307/278435

Butzer, K.W. (1964). Environment and Archaeology. Chicago: Aldine.

Candille, S.I., D.M. Absher, S. Beleza, M. Bauchet, B. McEvoy, N.A. Garrison, et al. (2012). Genome-wide association studies of quantitatively measured skin, hair, and eye pigmentation in four European populations. PLoS One 7(10): e48294.
https://doi.org/10.1371/journal.pone.0048294

Canfield, V.A., A. Berg, S. Peckins, S.M. Wentzel, K.C. Ang, S. Oppenheimer, and K.C. Cheng. (2014). Molecular phylogeography of a human autosomal skin color locus under natural selection. G3, 3(11): 2059-2067.
https://doi.org/10.1534/g3.113.007484

Capart, J. (1905). Primitive Art in Egypt. London: H. Grevel.

Carpenter, F. I. (1936). Puritans preferred blondes. The heroines of Melville and Hawthorne. New England Quarterly 9(2): 253-272.
https://doi.org/10.2307/360391

Carrito, M.L., I.M.B. dos Santos, C.E. Lefevre, R.D. Whitehead, C.F. da Silva, and D.I. Perrett. (2016). The role of sexually dimorphic skin colour and shape in attractiveness of male faces. Evolution and Human Behavior 37(2): 125-133.
https://doi.org/10.1016/j.evolhumbehav.2015.09.006

Cavalli-Sforza, L.L., P. Menozzi, and A. Piazza. (1994). The history and geography of human genes. Princeton: Princeton University Press.

Chadwick, N.K. (1966). The Druids. University of Wales Press.

Chagnon, N.A., M.V. Flinn, and T.F. Melancon. (1979). Sex-ratio variation among the Yanomamo Indians. In N.A. Chagnon and W. Irons (Eds.) Evolutionary Biology and Human Social Behavior: An Anthropological Perspective (pp. 290-320). North Scituate, Mass.: Duxbury Press.

Charles, C.A.D. (2003). Skin bleaching, self-hate, and Black identity in Jamaica. Journal of Black Studies 33(6): 711-728.
https://doi.org/10.1177%2F0021934703033006001

Chen, X., H. Chen, W. Cai, M. Maguire, B. Ya, F. Zuo, et al. (2017). The melanoma-linked "redhead" MC1R influences dopaminergic neuron survival. Annals of Neurology 81(3): 395-406.
https://doi.org/10.1002/ana.24852

Child, F.J. (1965). The English and Scottish Popular Ballads. New York: Cooper Square.

Chin, C.B.N. (2013). Cosmopolitan Sex Workers. Women and Migration in a Global City. Oxford: Oxford University Press.

Choi, S.C., and M. Trotter. (1970). Statistical study of the multivariate structure and race-sex differences of American White and Negro fetal skeletons. American Journal of Physical Anthropology 33(3): 307-312.
https://doi.org/10.1002/ajpa.1330330304

Clark, K.B. and M.P. Clark. (1947). Racial identification and preference in Negro children, in T.M. Newcomb and E.L. Hartley (Eds.) Readings in Social Psychology, (pp. 169-178), New York: Henry Holt.

Colby, A.M. (1965). The Portrait in Twelfth-Century French Literature. Geneva: Droz.

Cooper, J.S. (1971). New cuneiform parallels to the Song of Songs. Journal of Biblical Literature 90(2): 157-162.
https://doi.org/10.2307/3263757

Coplan, R., B. Coleman, and K. Rubin. (1998). Shyness and little boy blue: Iris pigmentation, gender, and social wariness in preschoolers. Developmental Psychobiology 32(1): 37-44.
https://doi.org/10.1002/(SICI)1098-2302(199801)32:1<37::AIDDEV4>3.0.CO;2-U

Cosquin, E. (1922). Les contes indiens et l'Occident. Paris: Librairie ancienne Honoré Champion.

Crothers, L.R., and M.E. Cummings. (2013). Warning signal brightness variation: Sexual selection may work under the radar of natural selection in populations of a polytypic poison frog. The American Naturalist 181(5): E116-E124. https://doi.org/10.1086/670010

Cruz, V. (2012). Age, gender and social class influences on skin colour preferences among Mozambican children. Journal of Psychology in Africa. 27(2): 139-142.
https://doi.org/10.1080/14330237.2012.10874532

Cunningham, M.R., A.R. Roberts, A.P. Barbee, P.B. Druen, and C-H. Wu. (1995). 'Their ideas of beauty are, on the whole, the same as ours': consistency and variability in the cross-cultural perception of female physical attractiveness. Journal of Personality and Social Psychology 68 (2): 261-279.
https://doi.org/10.1037/0022-3514.68.2.261.

Curry, W.C. (1916). The Middle English Ideal of Personal Beauty, as found in the Metrical Romances, Chronicles, and Legends of the XIII, XIV, and XV Centuries. Baltimore: J.H. Furst Co.

Czaplicka, M.A. (1969) Aboriginal Siberia. A Study in Social Anthropology. Clarendon.

Dahlberg, A.A., and V.M. Carbonell. (1961). The dentition of the Magdalenian female from Cap Blanc, France. Man 61: 49-50.
https://doi.org/10.2307/2795784

Dali, I.M. (2012). Problématique du phénotype. Approche comparative des esclavages dans la Tunisie du XIXe siècle. In R. Botte and A. Stella (Eds.) Couleurs de l'esclavage sur les deux rives de la Méditerranée (Moyen Âge-XXe siècle), (pp. 337-369). Paris: Éditions Karthala.
https://doi.org/10.4000/books.pur.123702

Danel, D.P., S. Wacewicz, Z. Lewandowski, P. Zywiczynski, and J.O. Perea-Garcia. (2018). Humans do not perceive conspecifics with a greater exposed sclera as more trustworthy: a preliminary cross-ethnic study of the function of the overexposed human sclera. Acta Ethologica 21: 203-208.
https://doi.org/10.1007/s10211-018-0296-5

Danel, D.P., S. Wacewicz, K. Kleisner, Z. Lewandowski, M.E. Kret, P. Zywiczynski, and J.O. Perea-Garcia. (2020). Sex differences in ocular morphology in Caucasian people: a dubious role of sexual selection in the

evolution of sexual dimorphism of the human eye. Behavioral Ecology and Sociobiology 74(115)
https://doi.org/10.1007/s00265-020-02894-1

Darwin, C. (1872). The Expression of the Emotions in Man and Animals. London: John Murray.

Darwin, C. (1936 [1888]). The Descent of Man and Selection in relation to Sex. reprint of 2nd edition, The Modern Library, New York: Random House.

Davis, M. D. (1990). Portrayals of women in prime-time network television: Some demographic characteristics. Sex Roles 23(5-6): 325-332.
http://psycnet.apa.org/doi/10.1007/BF00290052

Davis, R.C. (2004). Christian Slaves, Muslim Masters: White Slavery in the Mediterranean, the Barbary Coast, and Italy, 1500-1800. Basingstoke: Palgrave-Macmillan.
https://doi.org/10.1017/s0021853706221728

Dennis, L.K. (2009). Tanning behaviour among young frequent tanners is related to attitudes and not lack of knowledge about the dangers. Health Education Journal 68(3): 232-243
https://doi.org/10.1177/0017896909345195

Derbeneva, O.A., E.B. Starikovskaya, D.C. Wallace, and R.I. Sukernik, (2002). Traces of early Eurasians in the Mansi of Northwest Siberia revealed by mitochondrial DNA analysis. American Journal of Human Genetics 70: 1009-1014.
https://doi.org/10.1086/339524

Dexter, M.R. (1985) Indo-European reflection of virginity and autonomy, Mankind Quarterly 26(1): 57-74.

Diamond, J. (1994). Race without color. Discover (November) 15: 83-89.

Dickson, D.B. (1990). The dawn of belief. Religion in the upper Paleolithic of Southwestern Europe. Tucson: University of Arizona Press.

Ding, Q., Y. Hu, S. Xu, C.C. Wang, H. Li, R. Zhang, et al. (2014). Neanderthal origin of the haplotypes carrying the functional variant Val92Met in the MC1R in modern humans. Molecular Biology and Evolution 31(8): 1994-2003
https://doi.org/10.1093/molbev/msu180

Dixson B.J., A.F. Dixson, B. Morgan, and M.J. Anderson. (2006). Human physique and sexual attractiveness: Sexual preferences of men and women in Bakossiland, Cameroon. Archives of Sexual Behavior 36(3): 369-375. https://doi.org/10.1007/s10508-006-9093-8

Donnelly, M.P., P. Paschou, E. Grigorenko, D. Gurwitz, C. Barta, R-B. Lu, et al. (2012). A global view of the OCA2-HERC2 region and pigmentation. Human Genetics 131(5): 683-696. http://dx.doi.org/10.1007%2Fs00439-011-1110-x

Drummond, P.D. (1997). Correlates of facial flushing and pallor in anger - provoking situations. Personality and Individual Differences 23(4): 575-582. https://doi.org/10.1016/S0191-8869(97)00077-9

Duffy, D.L., N.F. Box, W. Chen, J.S. Palmer, G.W. Montgomery, M.R. James, N.K. Hayward, N.G. Martin, and R.A. Sturm. (2004). Interactive effects of MC1R and OCA2 on melanoma risk phenotypes. Human Molecular Genetics 13(4): 447-461. https://doi.org/10.1093/hmg/ddh043

Duffy, D.L., G.W. Montgomery, W. Chen, Z.Z. Zhao, L. Le, M.R. James, et al. (2007). A three-single-nucleotide polymorphism haplotype in intron 1 of OCA2 explains most human eye-color variation. American Journal of Human Genetics 80(2): 241-252. https://doi.org/10.1086/510885

Dumézil, G. (1970) Archaic Roman Religion, University of Chicago Press.

Dupanloup, I., L. Pereira, G. Bertorelle, F. Calafell, M.J. Prata,, A. Amorim, and G. Barbujani. (2003). A recent shift from polygyny to monogamy in humans is suggested by the analysis of worldwide Y-chromosome diversity. Journal of Molecular Evolution 57(1): 85-97. https://doi.org/10.1007/s00239-003-2458-x

Dupuis-Roy, N., S. Faghel-Soubeyrand, and F. Gosselin. (2019). Time course of the use of chromatic and achromatic facial information for sex categorization. Vision Research 157: 36-43. https://doi.org/10.1016/j.visres.2018.08.004

Dupuis-Roy, N., I. Fortin, D. Fiset, and F. Gosselin. (2009). Uncovering gender discrimination cues in a realistic setting. Journal of Vision 9(2): 10, 1-8. https://doi.org/10.1167/9.2.10

Eakley, A.L, and A.E. Houde. (2004). Possible role of female discrimination against 'redundant' males in the evolution of colour pattern polymorphism in

guppies. Proceedings of the Royal Society of London B: Biological Sciences 271(5): S299-S301.
http://doi.org/10.1098/rsbl.2004.0165

Eaverly, M.A. (2013). Tan Men/Pale Women. Color and Gender in Archaic Greece and Egypt, a Comparative Approach. Ann Arbor: The University of Michigan Press.
https://doi.org/10.3998/mpub.3080238

Edwards, E.A., and S.Q. Duntley. (1939). The pigments and color of living human skin. American Journal of Anatomy 65(1): 1-33.
https://doi.org/10.1002/aja.1000650102

Edwards, E.A., and S.Q. Duntley. (1949). Cutaneous vascular changes in women in reference to the menstrual cycle and ovariectomy. American Journal of Obstetrics & Gynecology 57(3): 501-509.
https://doi.org/10.1016/0002-9378(49)90235-5

Edwards, E.A., J.B. Hamilton, S.Q. Duntley, and G. Hubert. (1941). Cutaneous vascular and pigmentary changes in castrate and eunuchoid men. Endocrinology 28(1): 119-128.
https://doi.org/10.1210/endo-28-1-119

Eiberg, H., J. Troelsen, M. Nielsen, A. Mikkelsen, J. Mengel-From, K.W. Kjaer, and L. Hansen. (2008). Blue eye color in humans may be caused by a perfectly associated founder mutation in a regulatory element located within the HERC2 gene inhibiting OCA2 expression. Human Genetics 123(2): 177-187.
http://doi.org/10.1007/s00439-007-0460-x

Emlen, S.T., and L.W. Oring. (1977). Ecology, sexual selection, and the evolution of mating systems. Science 197(4300): 215-223.
http://dx.doi.org/10.1126/science.327542

Endler, J.A. (1980). Natural selection on color patterns in Poecilia reticulata. Evolution 34(1): 76-91.
https://doi.org/10.1111/j.1558-5646.1980.tb04790.x

Evans, S.T. (2000). Aztec royal pleasure parks: Conspicuous consumption and elite status rivalry. Studies in the History of Gardens & Designed Landscapes 20(3): 206-228.
https://doi.org/10.1080/14601176.2000.10435621

Evans-Pritchard, E.E. (1937). Witchcraft, Oracles, and Magic among the Azande. Oxford: Clarendon Press.

Farah, M. J. (1996). Is face recognition 'special?' Evidence from neuropsychology. Behavioural Brain Research 76(1-2): 181-189. https://doi.org/10.1016/0166-4328(95)00198-0

Farr, J.A. (1980). Social behavior patterns as determinants of reproductive success in the guppy Poecilia reticulata Peters (Pisces: Poeciliidae). Behaviour 74(1): 38-91.
http://doi.org/10.1163/156853980X00311

Fiala, V., P. Tureček, R.M. Akoko, Š. Pokorný, and K. Kleisner. (2022). Africans and Europeans differ in their facial perception of dominance and sex-typicality: a multidimensional Bayesian approach. Scientific Reports 12(1): 6821.
https://doi.org/10.1038/s41598-022-10646-6.

Fisher, A.W. (1972). Muscovy and the Black Sea slave trade. Canadian-American Slavic Studies 6 (4): 575-594.
https://doi.org/10.1163/221023972X00039

Fisher, A.W. (1973). Azov in the sixteenth and seventeenth centuries. Jahrbücher für Geschichte Osteuropas, Neue Folge 21 (2): 161-174.

Fitzgerald, F.S. (1992). The Great Gatsby. New York: Collier Books.

Flanagan, N., E. Healy, A. Ray, S. Philips, C. Todd, I.J. Jackson, et al. (2000). Pleiotropic effects of the melanocortin 1 receptor (MC1R) gene on human pigmentation. Human Molecular Genetics 9(17): 2531-2537.
https://doi.org/10.1093/hmg/9.17.2531

Flegr, J. (2010). Elastic, not plastic species: Frozen plasticity theory and the origin of adaptive evolution in sexually reproducing organisms. Biology Direct 5: 2.
https://doi.org/10.1186/1745-6150-5-2

Flegr, J., and K. Sýkorová. (2019). Skin fairness is a better predictor for impaired physical and mental health than hair redness. Scientific Reports 9: 18138.
https://doi.org/10.1038/s41598-019-54662-5

Fleure, H.J. (1945). The distribution of types of skin color. Geographical Review 35(4): 580-595.
https://doi.org/10.2307/210797

Fokuo, J. Konadu (2009). The lighter side of marriage: Skin bleaching in post-colonial Ghana. Institute of African Studies Research Review NS 25(1):47-66.
http://dx.doi.org/10.4314/rrias.v25i1.44933

Fornaciari, C. and F. Mallegni. (1987). Palaenutritional studies on skeletal remains of ancient populations from the Mediterranean area: An attempt to interpretation. Anthropologischer Anzeiger 45(4): 361-370.
https://www.jstor.org/stable/29539830

Fortes, M. (1945). The Dynamics of Clanship among the Tallensi. London: Oxford University Press.

Forti, I.R.N., and R.J. Young. (2016). Human commercial models' eye colour shows negative frequency-dependent selection. PLoS One 11(12) e0168458.
https://doi.org/10.1371/journal.pone.0168458

Franbourg, A., P. Hallegot, F. Baltenneck, C. Toutain, and F. Leroy. (2003). Current research on ethnic hair. Journal of the American Academy of Dermatology 48(6), supplement, S115-S119.
https://doi.org/10.1067/mjd.2003.277

French.CHINA.ORG.CA. (2016). L'Afrique envahie par 18 500 travailleurs du sexe chinois ? Du sensationnalisme pur et simple ! April 23
http://french.china.org.cn/foreign/txt/2016-04/23/content_38309565.htm

Friedrich, A. (1913). From the Congo to the Niger and the Nile. An account of the German Central African Expedition of 1910-1911 by Adolf Friedrich Duke of Mecklenburg, vol. 1. London: Duckworth and Co.
https://doi.org/10.5962/bhl.title.82427

Frost, P. (1988). Human skin color: A possible relationship between its sexual dimorphism and its social perception. Perspectives in Biology and Medicine 32(1): 38-58.
https://doi.org/10.1353/pbm.1988.0010

Frost, P. (1989). Human skin color: the sexual differentiation of its social perception. Mankind Quarterly 30: 3-16.
https://www.researchgate.net/publication/256306184_Human_skin_color_the_sexual_differentiation_of_its_social_perception

Frost, P. (1990). Fair women, dark men: the forgotten roots of colour prejudice. History of European Ideas 12(5): 669-79.
https://doi.org/10.1016/0191-6599(90)90178-H

Frost, P. (1994a). Geographic distribution of human skin color: A selective compromise between natural selection and sexual selection? Human Evolution 9(2): 141-153.
https://doi.org/10.1007/BF02437260

Frost, P. (1994b). Preference for darker faces in photographs at different phases of the menstrual cycle: Preliminary assessment of evidence for a hormonal relationship. Perceptual and Motor Skills 79(1): 507-14.
https://doi.org/10.2466/pms.1994.79.1.507

Frost, P. (2006). European hair and eye color - A case of frequency-dependent sexual selection? Evolution and Human Behavior 27(2): 85-103.
https://doi.org/10.1016/j.evolhumbehav.2005.07.002

Frost, P. (2007). Comment on Human skin-color sexual dimorphism: A test of the sexual selection hypothesis. American Journal of Physical Anthropology 133(1): 779-781.
https://doi.org/10.1002/ajpa.20555

Frost, P. (2008). Sexual selection and human geographic variation, Special Issue: Proceedings of the 2nd Annual Meeting of the NorthEastern Evolutionary Psychology Society. Journal of Social, Evolutionary, and Cultural Psychology 2(4): 169-191.
http://dx.doi.org/10.1037/h0099346

Frost, P. (2010). Femmes claires, hommes foncés. Les racines oubliées du colorisme. Quebec City: Les Presses de l'Université Laval, 202 p.
https://doi.org/10.7202/006424ar

Frost, P. (2011). Hue and luminosity of human skin: a visual cue for gender recognition and other mental tasks. Human Ethology Bulletin 26(2): 25-34.
https://www.researchgate.net/publication/256296588_Hue_and_luminosity_of_human_skin_a_visual_cue_for_gender_recognition_and_other_mental_tasks

Frost, P. (2014). The puzzle of European hair, eye, and skin color. Advances in Anthropology 4(2): 78-88.
https://doi.org/10.4236/aa.2014.42011

Frost, P. (2015). Evolution of long head hair in humans. Advances in Anthropology 5(4): 78-88.
https://doi.org/10.4236/aa.2015.54021

Frost, P. (2020a). An Accelerant of Social Change? The Spanish Flu of 1918-19. International Political Anthropology Journal 13(2): 123-133.
https://doi.org/10.5281/zenodo.4295574

Frost, P. (2020b). White Skin Privilege: Modern Myth, Forgotten Past. Evolutionary Studies in Imaginative Culture 4(2): 63-82.
https://doi.org/10.26613/esic/4.2.190

Frost, P., K. Kleisner, and J. Flegr. (2017). Health status by gender, hair color, and eye color: Red-haired women are the most divergent. PLoS One 12(12): e0190238.
https://doi.org/10.1371/journal.pone.0190238

Gao, X., K.C. Simon, J. Han, M.A. Schwarzchild, and A. Ascherio. (2009). Genetic determinants of hair color and Parkinson's disease risk. Annals of Neurology 65(1): 76-82.
https://doi.org/10.1002/ana.21535

Garza, R., R.R. Heredia, and A.B. Cieslicka. (2016). Male and female perception of physical attractiveness. An eye movement study. Evolutionary Psychology 14(1): 1-16.
https://doi.org/10.1177%2F1474704916631614

Gasperino, J. (1996). Ethnic differences in body composition and their relation to health and disease in women. Ethnicity & Health 1(4): 337-347.
https://doi.org/10.1080/13557858.1996.9961803

Gergen, K.J. (1967). The significance of skin color in human relations. Daedalus 96: 390-406.
http://www.jstor.org/stable/20027044

Gibbons, A. (2007). American Association of Physical Anthropologists Meeting: European skin turned pale only recently, gene suggests. Science 20 April 2007, 316(5823): 364.
https://doi.org/10.1126/science.316.5823.364a

Gimbutas, M.A. (1982) The Goddesses and Gods of Old Europe, 6,500-3,500 BC, myths and cult images. University of California.
https://doi.org/10.2307/2801144

Goebel, T. (1999). Pleistocene human colonization of Siberia and peopling of the Americas: An ecological approach. Evolutionary Anthropology 8(6): 208-227.
https://doi.org/10.1002/(SICI)1520-6505(1999)8:6<208::AID-EVAN2>3.0.CO;2-M

Goldenberg, D.M. (2003). The Curse of Ham. Race and Slavery in Early Judaism, Christianity, and Islam. Princeton: Princeton University Press.
https://doi.org/10.1515/9781400828548

Goodman, M.E. (1946). Evidence concerning the genesis of interracial attitudes. American Anthropologist 48(4): 624-630.
https://doi.org/10.1525/aa.1946.48.4.02a00080

Graf, K.E. (2009a). "The Good, the Bad, and the Ugly": evaluating the radiocarbon chronology of the middle and late Upper Paleolithic in the Enisei River valley, south-central Siberia. Journal of Archaeological Science 36(3): 694-707.
https://doi.org/10.1016/j.jas.2008.10.014

Graf, K.E. (2009b). Modern human colonization of the Siberian Mammoth Steppe: A view from South-Central Siberia. In M. Camps and P. Chauhan (Eds.) Sourcebook of Paleolithic Transitions, (pp. 484-496). Springer Science & Business Media.
https://doi.org/10.1007/978-0-387-76487-0_32

Grande, R., E. Gutierrez, E. Latorre, and F. Arguelles. (1994). Physiological variations in the pigmentation of newborn infants, Human Biology 66(3): 495-507.
http://www.jstor.org/stable/41465000

Grant, B., A. Snyder, and S.F. Glessner. (1974). Frequency-dependent mate selection in Mormoniella vitripennis. Evolution 28(2): 259-264.
https://doi.org/10.1111/j.1558-5646.1974.tb00745.x

Graves, R. and A. Hodge. (1950). The Long Week-end - A Social History of Great Britain 1918-1939. London: Faber.

Grimm, J. and W. Grimm. (1969). The Grimms' German Folk Tales. Carbondale: Southern Illinois University Press.

Gründl, M., S. Knoll, M. Eisenmann-Klein, and L. Pranti. (2012). The Blue-Eyes Stereotype: Do Eye Color, Pupil Diameter, and Scleral Color Affect Attractiveness? Aesthetic Plastic Surgery 36(2): 234-240.
https://doi.org/10.1007/s00266-011-9793-x

Guenther, C.A., B. Tasic, L. Luo, M.A. Bedell, and D.M. Kingsley. (2014). A molecular basis for classic blond hair color in Europeans. Nature Genetics 46(7): 748-752.
https://dx.doi.org/10.1038%2Fng.2991

Guilhem, D. (2006). Le lait de vache dans les sociétés peules. Pratiques alimentaires et symbolisme d'un critère identitaire. Partenariat Ocha/Les Cafés Géo, au Café Flore, Paris, April 25
http://www.mangeur-ocha.com/fileadmin/images/sciences_humaines/ Doroth_e_GUILHEM_Texte_Caf__G_o.pdf

Günther, T., H. Malmström, E.M. Svensson, A. Omrak, F. Sánchez-Quinto, G.M. Kilinç, et al. (2018). Population genomics of Mesolithic Scandinavia:

Investigating early postglacial migration routes and high-latitude adaptation. PLoS Biol 16(1): e2003703.
https://doi.org/10.1371/journal.pbio.2003703

Guthrie, R.D. (1970). Evolution of human threat display organs. In T. Dobzhansky, M.K. Hecht, and W.C. Steere (Eds.) Evolutionary Biology 4: 257-302. New York: Appleton-Century Crofts.

Haim, S.G. (1978). Love in an Arab Climate. Encounter 50: 86-91.

Hallowell, A.I. (1971). The Role of Conjuring in Saulteaux Society, Publications of the Philadelphia Anthropological Society, Vol. 2, Octagon.

Hampton, K.J., K.A. Hughes, and A.E. Houde. (2009). The allure of the distinctive: reduced sexual responsiveness of female guppies to 'redundant' male colour patterns. Ethology 115(5): 475-481.
https://doi.org/10.1111/j.1439-0310.2009.01634.x

Harding, R.M., E. Healy, A.J. Ray, N.S. Ellis, N. Flanagan, C. Todd, et al. (2000). Evidence for variable selective pressures at MC1R. American Journal of Human Genetics 66(4): 1351-1361.
https://doi.org/10.1086/302863

Harrold, F.B. (1980). A comparative analysis of Eurasian Paleolithic burials. World Archaeology 12(2): 195-211.
https://doi.org/10.1080/00438243.1980.9979792

Harvey, R. G. (1985). Ecological factors in skin color variation among Papua New Guineans, American Journal of Physical Anthropology 66(4): 407-416.
https://doi.org/10.1002/ajpa.1330660409

Healey, M., T. Uller, and M. Olsson. (2008). Variety is the spice of life: female lizards choose to associate with colour-polymorphic male dyads. Ethology 114(3): 231-237.
https://doi.org/10.1111/j.1439-0310.2007.01469.x

Heapost, L. (2008). Pigmentation variation in Finno-Ugric peoples. Papers on Anthropology XVII, pp. 80-102.

Henrich, J. (2017). The Secret of Our Success: How Culture is Driving Human Evolution, Domesticating Our Species, and Making Us Smarter. Princeton University Press.

Henry, J. (1910). L'âme d'un peuple africain. Münster: Aschendorff.

Higham, T., K. Douka, R. Wood, C. Bronk Ramsey, F. Brock, L. Basell, et al. (2014). The time and spatiotemporal patterning of Neanderthal disappearance. Nature 512(7514): 306-309.
https://doi.org/10.1038/nature13621

Hill, R., and R. Barton. (2005). Red enhances human performance in contests. Nature 435: 293.
https://doi.org/10.1038/435293a

Hill, H., V. Bruce, and S. Akamatsu. (1995). Perceiving the sex and race of faces: The role of shape and colour. Proceedings of the Royal Society B: Biological Sciences 261(1362): 367-373.
https://doi.org/10.1098/rspb.1995.0161

Hodge, F.W. ([1905]1959). Handbook of American Indians north of Mexico. Part 1. New York: Pageant.

Hoffecker, J.F. (2002). Desolate landscapes. Ice-age settlement in Eastern Europe. New Brunswick: Rutgers University Press.

Hogben, J.H., B. Julesz, and J. Ross. (1976). Short-term memory for symmetry. Vision Research 16(8): 861-866.
https://doi.org/10.1016/0042-6989(76)90146-2

Holy, L. (1988). Gender and ritual in an Islamic society: The Berti of Darfur. Man 23: 469-487.
https://doi.org/10.2307/2803261

Hönekopp, J. and S. Watson (2010). Meta-analysis of digit ratio 2D:4D shows greater sex difference in the right hand. American Journal of Human Biology 22(5): 619-630.
https://doi.org/10.1002/ajhb.21054

Howell, N. (1979). Demography of the Dobe !Kung. Academic Press.

Hrdlicka, A. (1898). Physical differences between White and Colored children. American Anthropologist 11(11): 347-350.
https://doi.org/10.1525/aa.1898.11.11.02a00020

Huang, X., S. Wang, L. Jin, and Y. He. (2021). Dissecting dynamics and differences of selective pressures in the evolution of human pigmentation. Biology Open 15 February 2021; 10(2): bio056523.
https://doi.org/10.1242/bio.056523

Hughes, K.A., L. Du, F.H. Rodd, and D.N. Reznick. (1999). Familiarity leads to female mate preference for novel males in the guppy, Poecilia reticulata. Animal Behavior 58(4): 907-916.
https://doi.org/10.1006/anbe.1999.1225

Hughes, K.A., F.H. Rodd, and D.N. Reznick. (2005). Genetic and environmental effects on secondary sex traits in guppies (Poecilia reticulata). Journal of Evolutionary Biology 18(1): 35-45.
https://doi.org/10.1111/j.1420-9101.2004.00806.x

Hulse, F.S. (1967). Selection for skin color among the Japanese. American Journal of Physical Anthropology 27(2): 143-156.
https://doi.org/10.1002/ajpa.1330270205

Hysi, P.G., A.M. Valdes, F. Liu, N.A. Furlotte, D.M. Evans, V. Bataille, et al. (2018). Genome-wide association meta-analysis of individuals of European ancestry identifies new loci explaining a substantial fraction of hair color variation and heritability. Nature Genetics 50(5): 652-656.
https://doi.org/10.1038/s41588-018-0100-5

Ikizler, A.S. (2007). Gender role representations in Turkish television programs, Submitted as a St. Mary's Project in Partial Fulfillment of the Graduation Requirements, St. Mary's College of Maryland for the Degree of Bachelor of Arts in Psychology
http://citeseerx.ist.psu.edu/viewdoc/summary?doi=10.1.1.514.2218

Ioan, S., M. Sandulache, S. Avramescu, A. Ilie, and A. Neacsu. (2007). Red is a distractor for men in competition. Evolution and Human Behavior 28: 285-293.
https://doi.org/10.1016/j.evolhumbehav.2007.03.001

Irwin, E. (1974). Colour Terms in Greek Poetry. Toronto: Hakkert.

Isaacs, H. R. (1967). Group identity and political change: The role of color and physical characteristics. Daedalus 96(2): 353-375.
https://www.jstor.org/stable/20027042

Iwawaki, S., K. Sonoo, J.E. Williams, and D.L. Best. (1978). Color bias among young Japanese children, Journal of Cross-Cultural Psychology 9(1): 61-73.
https://doi.org/10.1177/002202217891005

Jablonski, N.G., and G. Chaplin. (2000). The evolution of human skin coloration. Journal of Human Evolution 39(1): 57-106, p. 71, 76.
https://doi.org/10.1006/jhev.2000.0403

Janif, Z.J., R.C. Brooks, and B.J. Dixson. (2015). Are preferences for women's hair color frequency-dependent? Adaptive Human Behavior and Physiology 1(1): 54-71.
https://doi.org/10.1007/s40750-014-0008-y

Jaquet, E., and G. Rhodes. (2008). Face aftereffects indicate dissociable, but not distinct, coding of male and female faces. Journal of Experimental Psychology: Human Perception and Performance 34(1): 101-112.
https://doi.org/10.1037/0096-1523.34.1.101

Jay, P.C. (1962). Aspects of maternal behavior among langurs, Annals of the New York Academy of Sciences 102(2): 468-476.
https://doi.org/10.1111/j.1749-6632.1962.tb13653.x

Jha, S. and M. Adelman. (2009). Looking for love in all the white places: a study of skin color preferences on Indian matrimonial and mate-seeking websites. Studies in South Asian Film and Media 1(1): 65-83.
https://doi.org/10.1386/safm.1.1.65/1

Jiemin, B. (2003). The gendered biopolitics of marriage and immigration: A study of pre-1949 Chinese immigrants in Thailand. Journal of Southeast Asian Studies, 34(1): 127-151.
https://doi.org/10.1017/S0022463403000079

Jonassen, C.T. (1951). Some historical and theoretical bases of racism in northwestern Europe. Social Forces 30(2): 155-161.
https://doi.org/10.2307/2571627

Jones, A.L., R. Russell, and R. Ward. (2015). Cosmetics alter biologically-based factors of beauty: evidence from facial contrast. Evolutionary Psychology 13(1):
https://doi.org/10.1177%2F147470491501300113

Jones, E.R., G. Zarina, V. Moiseyev, E. Lightfoot, P.R. Nigst, A. Manica, et al. (2017). The Neolithic transition in the Baltic was not driven by admixture with early European farmers, Current Biology 27(4): 576-582.
https://doi.org/10.1016/j.cub.2016.12.060

Kahlon, D. P. S. (1976). Age variation in skin color, a study in Sikh immigrants in Britain, Human Biology 48(3): 419-428.

Kalla, A.K. (1973). Ageing and sex differences in human skin pigmentation. Zeitschrift für Morphologie und Anthropologie 65(1): 29-33.
http://www.jstor.org/stable/25756080

Kalla, A. K. and Tiwari, S.C. (1970). Sex differences in skin colour in man, Acta Geneticae Medicae et Gemellologiae 19(3): 472-476. https://doi.org/10.1017/S1120962300014876

Kamberov, Y.G., S. Wang, J. Tan, P. Gerbault, A. Wark, L. Tan, et al. (2013). Modeling Recent Human Evolution in Mice by Expression of a Selected EDAR Variant. Cell 152(4): 691-702. http://dx.doi.org/10.1016/j.cell.2013.01.016

Karras, R.M. (1988). Slavery and Society in Medieval Scandinavia. New Haven.

Kashuba, N. (2012). The neolithization of Ukraine. Bachelor thesis, Archaeology, Stockholm University.

Katsara, M-A., and M. Nothnagel. (2019). True colors: A literature review on the spatial distribution of eye and hair pigmentation. Forensic Science International: Genetics 39: 109-118. https://doi.org/10.1016/j.fsigen.2019.01.001

Kayser, M., F. Liu, A.C.J.W. Janssens, F. Rivadeneira, O. Lao, K. van Duijn, et al.. (2008). Three genome-wide association studies and a linkage analysis identify HERC2 as a human iris color gene. American Journal of Human Genetics 82(2): 411-423. https://doi.org/10.1016/j.ajhg.2007.10.003

Kažoka, D. and J. Vetra. (2011). Variations in some anthropometrical parameters of the women with the different iris color in Latvia. Papers on Anthropology XX: 160-170. https://doi.org/10.12697/poa.2011.20.17

Keane, A.H. (1886). Asia with Ethnological Appendix. London: Edward Stanford.

Keeler, C. (1970). Cuna moon-child albinism, 1950-1970. Journal of Heredity 61(6): 273-278. https://doi.org/10.1093/oxfordjournals.jhered.a108100

Kelly, R.L. (1995). The foraging spectrum. Diversity in hunter-gatherer lifeways. Washington: Smithsonian Institution Press.

Kenny, E.E., N.J. Timpson, M. Sikora, M-C. Yee, A. Moreno-Estrada, C. Eng, S. Huntsman, E.G. Burchard, M. Stoneking, C.D. Bustamante, and S. Myles (2012). Melanesian blond hair is caused by an amino acid change in TYRP1. Science 336(6081): 554. http://dx.doi.org/10.1126/science.1217849

Khumalo, N.P. (2005). African hair morphology: macrostructure to ultrastructure. International Journal of Dermatology 44(Suppl. 1): 10-12. https://doi.org/10.1111/j.1365-4632.2005.02805.x

Khumalo, N.P. (2008). On the history of African hair care: more treasures await discovery. Letter to the Editor. Journal of Cosmetic Dermatology, 7(3): 231. https://doi.org/10.1111/j.1473-2165.2008.00396.x

King, M-C, and A.C. Wilson. (1975). Evolution at two levels in humans and chimpanzees. Science 188(4184): 107-116. https://doi.org/10.1126/science.1090005

Kirkpatrick, M. (1987). Sexual selection by female choice in polygynous animals. Annual Review of Ecology and Systematics 18: 43-70. https://doi.org/10.1146/annurev.es.18.110187.000355

Kisilov, M. (2007). Slave Trade in the Early Modern Crimea from the Perspective of Christian, Muslim, and Jewish Sources. Journal of Early Modern History 11(1-2): 1-32. https://doi.org/10.1163/157006507780385125

Kjellström, R. (1973). Eskimo marriage. An account of traditional Eskimo courtship and marriage. Lund: Nordiska Museets Handlingar 80. https://doi.org/10.2307/2800119

Kleisner, K., T. Kocnar, A. Rubešová, and J. Flegr. (2010). Eye color predicts but does not directly influence perceived dominance in men. Personality and Individual Differences 49(1): 59-64. https://doi.org/10.1016/j.paid.2010.03.011

Kleisner, K., T. Kocnar, P. Turecek, D. Stella, R.M. Akoko, V. Trebický, and J. Havlícek. (2017). African and European perception of African female attractiveness. Evolution of Human Behavior 38(6): 744-755. https://doi.org/10.1016/j.evolhumbehav.2017.07.002

Kleisner, K., L. Priplatova, P. Frost, and J. Flegr. (2013). Trustworthy-looking face meets brown eyes. PLoS One 8(1): e53285. https://doi.org/10.1371/journal.pone.0053285

Kligman, G., and S. Limoncelli. (2005). Trafficking women after socialism: To, through, and from Eastern Europe. Social Politics: International Studies in Gender, State and Society 12(1): 118-140. https://doi.org/10.1093/sp/jxi006

Kokko, H., M.D. Jennions, and A. Houde. (2007). Evolution of frequency-dependent mate choice: keeping up with fashion trends. Proceedings of the Royal Society of London B: Biological Sciences 274(1615): 1317-1324. http://dx.doi.org/10.1098/rspb.2007.0043

Kolodziejczyk, D. (2006). Slave hunting and slave redemption as a business enterprise: The northern Black Sea region in the sixteenth to seventeenth centuries. Oriente Moderno 86 (1): 149-159.

Kondo, S., Y. Hozumi, and K. Aso. (1990). Organ culture of human scalp hair follicles: effect of testosterone and oestrogen on hair growth, Archives of Dermatological Research 282(7): 442-445. https://doi.org/10.1007/BF00402619

Koslowski. (2013). Global Human Smuggling: Comparative Perspectives, Baltimore: The Johns Hopkins University Press. https://doi.org/10.1007/s12142-015-0364-4

Kozlowski, S.K., and E. Sachse-Kozlowska. (1995). Magdalenian family from the Maszycka Cave. Jahrbuch der Römisch Germanischen Zentral Museums Mainz 40(1): 115-205.

Krupnik, I.I. (1985). The male-female ratio in certain traditional populations of the Siberian Arctic. Inuit Studies 9(1): 115-140. http://www.jstor.org/stable/42869442

Kuraguchi, K. and K. Kanari. (2021). Enlargement of female pupils when perceiving something cute. Scientific Reports 11: 23367 https://doi.org/10.1038/s41598-021-02852-5

Lambert, W.G. (1959). Divine love lyrics from Babylon. Journal of Semitic Studies 4(1): 1-15. https://doi.org/10.1093/jss/4.1.1

Langlois, J.H., L. Kalakanis, A.J. Rubenstein, A. Larson, M. Hallam, and M. Smoot. (2000). Maxims or myths of beauty? A meta-analytic and theoretical review. Psychological Bulletin 126(3): 390-423. https://doi.org/10.1037//0033-2909.126.3.390.

Langlois, J.H., J.M. Ritter, L.A. Roggman, and L.S. Vaughn. (1991). Facial diversity and infant preferences for attractive faces. Developmental Psychology 27(1): 79-84. https://doi.org/10.1037/0012-1649.27.1.79.

Langlois, J.H., L.A. Roggman, R.J. Casey, and J.M. Ritter. (1987). Infant preferences for attractive faces: Rudiments of a stereotype? Developmental Psychology 23(3): 363-369.
https://doi.org/10.1037//0012-1649.23.3.363.

Langlois, J.H. and C. Stephan. (1977). The effects of physical attractiveness and ethnicity on children's behavioral attributions and peer preferences. Child Development 48(4): 1694-1698.
https://doi.org/10.2307/1128538.

Lank, D.B. (2002). Diverse processes maintain plumage polymorphisms in birds. Journal of Avian Biology 33(4): 327-330.
https://doi.org/10.1034/j.1600-048X.2002.30811.x

Lao, O., J.M. de Gruijter, K. van Duijn, A. Navarro, and M. Kayser. (2007). Signatures of positive selection in genes associated with human skin pigmentation as revealed from analyses of single nucleotide polymorphisms. Annals of Human Genetics 71(3): 354–369.
https://doi.org/10.1111/j.1469-1809.2006.00341.x

Lawrence, W. (1848). Lectures on Comparative Anatomy, Physiology, Zoology, and the Natural History of Man. London: Henry G. Bohn.

Lazaridis, I., D. Nadel, G. Rollefson, D.C. Merrett, N. Rohland, S. Mallick, et al. (2016). Genomic insights into the origin of farming in the ancient Near East. Nature 536: 419-424.
http://dx.doi.org/10.1038/nature19310

Lazaridis, I., N. Patterson, A. Mittnik, G. Renaud, S. Mallick, K. Kirsanow, et al. (2014). Ancient human genomes suggest three ancestral populations for present-day Europeans. Nature 513(7518): 409-413.
https://doi.org/10.1038/nature13673

Lea-Greenwood, G. (1998). Visual merchandising: a neglected area in UK fashion marketing? International Journal of Retail & Distribution Management 26(8): 324-329.
https://doi.org/10.1108/09590559810231797

League of Nations. (1934). Commission of Enquiry into Traffic in Women and Children in the East, Summary of the Report to the Council, IV, Social 1934.IV 3, cited in D. Kyle and R. Koslowski (2013). Global Human Smuggling: Comparative Perspectives. Baltimore: The Johns Hopkins University Press.
https://doi.org/10.1007/s12142-015-0364-4.

Lee, T.J., M.P. Speed, and P.A. Stephens. (2011). Honest signaling and the uses of prey coloration. The American Naturalist 178(1): E1-E9. https://doi.org/10.1086/660197

Lefèvre, A. (1900) Les Gaulois - origines et croyances, Librairie C. Reinwald.

Lepowsky, M. (1987). Food taboos and child survival: A case study from the Coral Sea, in N. Scheper-Hughes (Ed.) Child Survival: Anthropological Perspectives on the Treatment and Maltreatment of Children (pp. 71-92), Springer. https://doi.org/10.1007/978-94-009-3393-4_4

Lévi-Provençal, É. (1953). Histoire de l'Espagne musulmane, tome III. Paris: G.P. Maisonneuve.

Lewis, B. (1990). Race and Slavery in the Middle East. New York: Oxford University Press.

Lewis, I.M. (1962). Marriage and the Family in Northern Somaliland. Kampala: East African Institute of Social Research.

Liberton, D.K., K.A. Matthes, R. Pereira, T. Frudakis, D.A. Puts, and M.D. Shriver. (2009). Patterns of correlation between genetic ancestry and facial features suggest selection on females is driving differentiation, Poster #326. American Society of Human Genetics, 59th annual meeting, October 20-24. Honolulu, Hawaii.

Liem, E.B., S.C. Hollensead, R. Lenhardt, and D.I. Sessler. (2003). Decreased platelet function in people with red hair. Anesthesiology. 99(3A): A146.

Liem, E.B., T.V. Joiner, K. Tsueda, and D.I. Sessler. (2005). Increased sensitivity to thermal pain and reduced subcutaneous lidocaine efficacy in redheads. Anesthesiology. 102(3): 509-514. https://doi.org/10.1097/00000542-200503000-00006

Little, A.C., L.M. DeBruine, and B.C. Jones. (2005). Sex-contingent face after-effects suggest distinct neural populations code male and female faces. Proceedings of the Royal Society of London, Series B 272(1578): 2283-2287. https://doi.org/10.1098/rspb.2005.3220

Liu, F., Y. Chen, G. Zhu, P.G. Hysi, S. Wu, K. Adhikari. (2018). Meta-analysis of genome-wide association studies identifies 8 novel loci involved in shape variation of human head hair. Human Molecular Genetics 27(3): 559-575. https://doi.org/10.1093/hmg/ddx416

Liu, F., A. Wollstein, P.G. Hysi, G.A. Ankra-Badu, T.D. Spector, D. Park, et al. (2010). Digital quantification of human eye color highlights genetic association of three new loci. PLoS Genetics 6(5): e1000934.
https://doi.org/10.1371/journal.pgen.1000934

Lorenz, K. (1937). The companion in the bird's world. The Auk 54(3): 245-273.
https://doi.org/10.2307/4078077

Lorenz, K. (1971). Studies in Animal and Human Behaviour, vol. 2. London: Methuen & Co.

Loussouarn, G. (2001). African hair growth parameters. British Journal of Dermatology 145(2): 294-297.
https://doi.org/10.1046/j.1365-2133.2001.04350.x

Loussouarn, G., C.E. Rawadi, and Genain, G. (2005). Diversity of hair growth profiles. International Journal of Dermatology 44(Suppl. 1), 6-9.
https://doi.org/10.1111/j.1365-4632.2005.02800.x

Lugira, A.M. (1970). Ganda Art. Kampala: Osasa pub.

Maan, M.E., and M.E. Cummings. (2009). Sexual dimorphism and directional sexual selection on aposematic signals in a poison frog. Proceedings of the National Academy of Sciences USA 106(45): 19072-19077.
https://doi.org/10.1073/pnas.0903327106

Maanasa, R., P. Skoglund, K.E. Graf, M. Metspalu, A. Albrechtsen, I. Moltke, et al. (2013). Upper Palaeolithic Siberian genome reveals dual ancestry of Native Americans. Nature 505(7481): 87-91.
https://doi.org/10.1038/nature12736

Macintosh, N.W.G. (1959). A preliminary note on skin colour in the western highland natives of New Guinea, Oceania 30(4): 279-293.
https://doi.org/10.1002/j.1834-4461.1960.tb00227.x

Madrigal, L., and W. Kelly. (2006). Human skin-color sexual dimorphism: A test of the sexual selection hypothesis. American Journal of Physical Anthropology 132(3): 470-482.
https://doi.org/10.1002/ajpa.20453

Maertens, J-T. (1978). Le dessein sur la peau. Essai d'anthropologie des inscriptions tégumentaires, Ritologiques I, Paris: Aubier Montaigne.
https://doi.org/10.7202/000904ar

Mageo, J.M. (1994). Hairdos and Don'ts: Hair Symbolism and Sexual History in Samoa. Man, New Series, 29(2): 407-432. https://doi.org/10.2307/2804480

Majid, A., and N. Kruspe. (2018). Hunter-gatherer olfaction is special. Current Biology 28: R108-R110. https://doi.org/10.1016/j.cub.2017.12.014

Makova, K., and H. Norton. (2005). Worldwide polymorphism at the MC1R locus and normal pigmentation variation in humans. Peptides 26(10): 1901-1908. https://doi.org/10.1016/j.peptides.2004.12.032

Malinowski, B. (1932). The Sexual Life of Savages in North-Western Melanesia. London: Routledge & Kegan Paul.

Mallegni, F., and P.F. Fabbri. (1995). The human skeletal remains from the Upper Paleolithic burials found in Romito Cave (Papasidero, Cosenza, Italy). Bulletin et Mémoires de la Société d'Anthropologie de Paris, n.s 7(3-4): 99-137. https://doi.org/10.3406/bmsap.1995.2413

Manning, A. (1972). An Introduction to Animal Behaviour. 2nd edition. London: Edward Arnold.

Manning, J.T., P.E. Bundred, and F.M. Mather. (2004). Second to fourth digit ratio, sexual selection, and skin colour. Evolution and Human Behavior 25(1): 38-50. https://doi.org/10.1016/s1090-5138(03)00082-5

Manning, P. (1983). Contours of slavery and social change in Africa. The American Historical Review 88 (4): 835-857. https://doi.org/10.2307/1874022

Maret, S.M. (1983). Attractiveness ratings of photographs of Blacks by Cruzans and Americans. The Journal of Psychology 115 (1): 113-116. https://doi.org/10.1080/00223980.1983.9923605.

Marks, R., P.A. Foley, and D. Jolley. (1995). The effect of regular sunscreen use on vitamin D levels in an Australian population. Results of a randomized control trial. Archives of Dermatology 131(4): 415-421. https://doi.org/10.1001/archderm.1995.01690160043006

Martin, M.K. (1974). The foraging adaptation - Uniformity or diversity? Addison-Wesley Module in Anthropology 56.

Martinengo-Cesaresco, E. (1886). Essays in the Study of Folk-Songs. London: J.M. Dent & Sons.

Martinez-Cadenas, C., M. Pena-Chilet, M. Ibarrola-Villava, and G. Ribas. (2013). Gender is a major factor explaining discrepancies in eye colour prediction based on HERC2/OCA2 genotype and the IrisPlex model. Forensic Science International: Genetics 7(4): 453-460.
https://doi.org/10.1016/j.fsigen.2013.03.007

Mason, S.L.R., J.G. Hather, and G.C. Hillman. (1994). Preliminary investigation of the plant macro-remains from Dolní Vestonice II, and its implications for the role of plant foods in Palaeolithic and Mesolithic Europe. Antiquity 68(258): 48-57.
https://doi.org/10.1017/S0003598X00046184

Massny. D. (1937). Die Formel "das braune Mägdelein" im alten deutschen Volkslied. Niederdeutsche Zeitschrift für Volkskunde 15: 26-65.

Mathieson, I, I. Lazaridis, N. Rohland, S. Mallick, N. Patterson, S. Alpaslan, et al. (2015). Genome-wide patterns of selection in 230 ancient Eurasians. Nature 528(7583): 499-503.
https://doi.org/10.1038/nature16152

Mathieson, I., S. Alpaslan-Roodenberg, C. Posth, A. Szécsényi-Nagy, N. Rohland, S. Mallick, et al. (2018). The Genomic History of Southeastern Europe, Supplementary Information, p. 52. Nature 555: 197-203.
http://dx.doi.org/10.1038/nature25778

Mazess, R.B. (1967). Skin color in Bahamian Negroes. Human Biology 39(2): 145-154.

McGovern, J.R. (1968). The American woman's pre-World War I freedom in manners and morals. Journal of American History 55(2): 315-333.
https://www.jstor.org/stable/1899561

McGuiness, B.W. (1961). Skin pigmentation and the menstrual cycle. British Medical Journal 2(5251): 563-565.
http://www.jstor.org/stable/20354588

McVicar, K.G. (1969). Twilight of an East African Slum. Ann Arbor, University Microfilms (UCLA Dissertation 1968).

Mead, W.E. (1969). "Color in the English and Scottish popular ballads," in F.J. Furnivall (Ed.). An English Miscellany, (pp. 321-334), New York: Benjamin Blom.

Medland, S.E., D.R. Nyholt, J.N. Painter, B.P. McEvoy, A.F. McRae, G. Zhu, et al. (2009). Common variants in the Trichohyalin gene are associated with straight hair in Europeans. American Journal of Human Genetics 85(5): 750-755. https://doi.org/10.1016/j.ajhg.2009.10.009

Melchior, L., N. Lynnerup, H.R. Siegismund, T. Kivisild, and J. Dissing. (2010). Genetic diversity among ancient Nordic populations. PLoS One 5(7): e11898 https://doi.org/10.1371/journal.pone.0011898

Meredith, H.V., and J.H. Spurgeon. (1980). Somatic comparisons at age 9 years for South Carolina White Girls and girls of other ethnic groups. Human Biology 52(3): 401-411.

Merker, M. (1910). Die Masai. Berlin: Reimer.

Mesa, M.S. (1983). Analyse de la variabilité de la pigmentation de la peau durant la croissance, Bulletin et mémoires de la Société d'Anthropologie de Paris, t. 10 série 13, 49-60.
https://doi.org/10.3406/bmsap.1983.3882

Miller, E.L. (1969). Body image, physical beauty and colour among Jamaican adolescents. Social and Economic Studies 18 (1): 72-89.

Millingen, F. (1870-1871). The Circassian slaves and the sultan's harem. Journal of the Anthropological Society of London 8: cix-cxx.
https://doi.org/10.2307/3025176

Missmer, S.A., D. Spiegelman, S.E. Hankinson, S. Malspeis, R.L. Barbieri, and D.J. Hunter. (2006). Natural hair color and the incidence of endometriosis. Fertility and Sterility 85(4): 866-870.
https://doi.org/10.1016/j.fertnstert.2005.12.008

Mittnik, A., C-C. Wang, S. Pfrengle, M. Daubaras, G. Zarina, F. Hallgren, et al. (2018). The genetic prehistory of the Baltic Sea region. Nature Communications 9(442)
https://doi.org/10.1038/s41467-018-02825-9

Montiel-Sosa, F., E. Ruiz-Pesini, J.A. Enriquez, A. Marcuello, C. Diez-Sanchez, J. Montoya, D.J. Wallace, and M.J. López-Pérez, (2006). Differences of sperm motility in mitochondrial DNA haplogroup U sublineages. Gene 368: 21-27.
https://doi.org/10.1016/j.gene.2005.09.015

Morgan, M.D., E. Pairo-Castineira, K. Rawlik, O. Canela-Xandri, J. Rees, D. Sims, A. Tenesa, and I.J. Jackson. (2018). Genome-wide study of hair colour in

UK Biobank explains most of the SNP heritability. Nature Communications 9: 5271
https://doi.org/10.1038/s41467-018-07691-z

Muggleton, J. (1979). Non-random mating in wild populations of polymorphic Adalia Bipunctata. Heredity 42: 57-65.
https://doi.org/10.1038/hdy.1979.6

Muhawi, I. and S. Kanaana. (1989). Speak, Bird, Speak Again. Berkeley: University of California Press.
https://doi.org/10.1525/california/9780520062924.001.0001

Munitz, S., B. Priel, and A. Henik. (1987). Color, skin color preferences and self color identification among Ethiopian and Israeli born children. In M. Ashkenazi and A. Weingrod (Eds.) Ethiopian Jews and Israel, (pp. 74-84), New Brunswick (U.S.A.): Transaction Books.

Ndjio, B. (2009). 'Shanghai beauties' and African desires: migration, trade and Chinese prostitution in Cameroon. The European Journal of Development Research 21(4): 606-621.
https://doi.org/10.1057/ejdr.2009.31

Ndjio, B. (2014a). "Magic body" and "cursed sex": Chinese sex workers as "bitch-witches" in Cameroon. African Affairs 113(452): 370-386.
https://doi.org/10.1093/afraf/adu042

Ndjio, B. (2014b). Sino-pessimism on the rise in Cameroon as Chinese sex labour migrants gain popularity. Africa at LSE, October 27, Firoz Lalji Centre for Africa https://blogs.lse.ac.uk/africaatlse/2014/10/27/sino-pessimism-on-the-rise-in-cameroon-as-chinese-sex-labour-migrants-gain-popularity/

Nelson, J.K., and K.R. Nelson. (1986). Skinfold profiles of Black and White boys and girls ages 11-13. Human Biology 58(3): 379-390.

Nestor, A., and M.J. Tarr. (2008a). The segmental structure of faces and its use in gender recognition. Journal of Vision, 8(7): 7, 1-12, http://journalofvision.org/8/7/7/, https://doi.org/10.1167/8.7.7.

Nestor, A., and M.J. Tarr. (2008b). Gender recognition of human faces using color. Psychological Science 19(12): 1242-1246.
https://doi.org/10.1111/j.1467-9280.2008.02232.x

Neto, F., and J.E. Williams. (1997). Color bias in children revisited: Findings from Portugal. Social Behavior and Personality: An International Journal 25(2): 115-121.
https://doi.org/10.2224/sbp.1997.25.2.115

Nickerson, E.C. (1926). Nature's Cosmetics, Bulletin sanitaire 26(5): 134-140.

Nordenskiold, A.E. (1882). The voyage of the Vega round Asia and Europe, with a historical review of previous journeys along the north coast of the old world. New York: Macmillan & Co.
https://doi.org/10.1017/cbo9781139235945.012

Norton, H.L., J.S. Friedlaender, D.A. Merriwether, G. Koki, C.S. Mgone, and M.D. Shriver. (2006). Skin and Hair Pigmentation Variation in Island Melanesia. American Journal of Physical Anthropology 130(2): 254-268.
https://doi.org/10.1002/ajpa.20343

Norton, H.L., and M.F. Hammer. (2007). Sequence variation in the pigmentation candidate gene SLC24A5 and evidence for independent evolution of light skin in European and East Asian populations. Program of the 77th Annual Meeting of the American Association of Physical Anthropologists, p. 179.

Nyborg, H. (1994). Hormones, Sex, and Society. The Science of Physicology. Westport, Conn.: Praeger.

Oakley, M. (ed.) (1990). Design management. A handbook of issues and methods. Oxford: Basil Blackwell.

Ogumefu, M.I. (1929). Yoruba Legends. London: The Sheldon Press.

Olalde, I., M.E. Allentoft, F. Sanchez-Quinto, G. Santpere, C.W.K. Chiang, M. DeGiorgio, et al. (2014). Derived immune and ancestral pigmentation alleles in a 7,000-year-old Mesolithic European. Nature 507 (7491): 225-228.
https://doi.org/10.1038/nature12960

Olalde, I. and C. Lalueza-Fox. (2015). Modern humans' paleogenomics and the new evidences on the European prehistory. Science and Technology of Archaeology Research 1(1): 1-9.
https://doi.org/10.1179/2054892315Y.0000000002

Olendorf, R., F.H. Rodd, D. Punzalan, A.E. Houde, C. Hurt, D.N. Reznick, and K.A. Hughes. (2006). Frequency-dependent survival in natural guppy populations. Nature 441(7093): 633-636.
https://doi.org/10.1038/nature04646

Olivier, G. (1960). Pratique anthropologique. Paris: Vigot Frères.

Omoto, K. (1965). Measurements of skin reflectance in a Japanese twin sample, Journal of the Anthropological Society of Nippon (Jinruigaku Zassi) 73(4): 115-122.
https://doi.org/10.1537/ase1911.73.115

Ondongo, J. (1984). Noir ou blanc ? Le vécu du double dans la pratique du “maquillage” chez les Noirs. Nouvelle Revue d’Ethnopsychiatrie 2:37-65.

O Regan, K. (2014). Red hair in popular culture and the relationship with anxiety and depression. B.A. (Honours) thesis, Psychology, University College Cork, Ireland.

Phillips, W.D. (1985). Slavery from Roman Times to the Early Transatlantic Trade. Manchester: Manchester University Press.

Osgood, C.E. (1960). The cross-cultural generality of visual-verbal synesthetic tendencies. Behavioral Science 5(2): 146-169.
https://doi.org/10.1002/bs.3830050204

Ossip-Drahos, A.G., J.R.O. Morales, C. Vital-Garcia, J.J. Zúñiga-Vega, D.K. Hews, and E.P. Martins. (2016). Shaping communicative colour signals over evolutionary time. Royal Society Open Science 3(11)
https://doi.org/10.1098/rsos.160728

Pallottino, M. (1952). Etruscan Painting. Lausanne: Skira.

Parizkova, J. (1977). Body Fat and Physical Fitness. The Hague: Martinus Nijhoff.

Pascalis, O., and D.J. Kelly. (2008). Face processing, in M. Haith and J. Benson (Eds.) Encyclopedia of Infant and Early Childhood Development, (pp. 471-478), Elsevier.
https://doi.org/10.1016/b978-012370877-9.00059-1

Perrett, D.I., K.A. May, and S. Yoshikawa. (1994). Facial shape and judgements of female attractiveness. Nature 368 (6468): 239-242.
https://doi.org/10.1038/368239a0.

Perrett, D.I., and R. Sprengelmeyer. (2021). Clothing aesthetics: consistent colour choices to match fair and tanned skin tones. i-Perception12(6): 204166952110533
http://dx.doi.org/10.1177/20416695211053361

Pincott, J. (2008). Do Gentlemen really prefer blondes? Bodies, Behavior, and Brains. The Science behind Sex, Love, and Attraction. New York: Delacorte Press.

Pospiech, E., J. Karlowska-Pik, M. Marcinska, S. Abidi, J. Dyrberg Andersen, M. van den Berge, et al. (2015). Evaluation of the predictive capacity of DNA variants associated with straight hair in Europeans. Forensic Science International: Genetics 19: 280-288.
https://doi.org/10.1016/j.fsigen.2015.09.004

Post, P.W., A.N. Krauss, S. Waldman, and Peter A.M. Auld. (1976). Skin reflectance of newborn infants from 25 to 44 weeks gestational age. Human Biology 48(3): 541-557.
http://www.jstor.org/stable/41462904

Powdermaker, H. (1956). Social change through imagery and values of teen-age Africans in Northern Rhodesia. American Anthropologist 58(5): 783-813.
https://doi.org/10.1525/aa.1956.58.5.02a00030

Price, T.D. (1991). The Mesolithic of Northern Europe. Annual Review of Anthropology 20: 211-233.
https://doi.org/10.1146/annurev.an.20.100191.001235

Proudhon, Pierre-Joseph. (1865). Du principe de l'art et de sa destination sociale. Paris : Garnier Frères.
https://doi.org/10.4000/books.pum.16219

Rana, B.K., D. Hewett-Emmett, L. Jin, B.H.J. Chang, N. Sambuughin, M. Lin, et al. (1999). High polymorphism at the human melanocortin 1 receptor locus. Genetics 151(4): 1547-1557.

Ray, N., and J.M. Adams. (2001). A GIS-based vegetation map of the world at the last glacial maximum (25,000-15,000 BP)-(23rd millennium to 13th millennium BC). Internet Archaeology 11.
https://doi.org/10.11141/ia.11.2

Read, S. (2012). "Only kept up by the credulous and ignorant": Eighteenth-century responses to the ancient beliefs about menstrual blood. Loughborough University. Chapter.
https://hdl.handle.net/2134/10058327.v1

Reich, D. (2018). Who We Are and How We Got There. Ancient DNA and the New Science of the Human Past. New York: Pantheon Books.
https://doi.org/10.1111/padr.12196

Renault, F. and S. Daget. (1985). Les traites négrières en Afrique. Paris: Éditions Karthala.

Renninger, C.A. and J.E. Williams. (1966). Black-white color connotations and racial awareness in preschool children. Perceptual and Motor Skills, 22(3): 771-785.
https://doi.org/10.2466/pms.1966.22.3.771

Richards, M.P., R.E.M. Hedges, R. Jacobi, A. Current, and C. Stringer. (2000). FOCUS: Gough's cave and sun hole cave human stable isotope values indicate a high animal protein diet in the British Upper Paleolithic. Journal of Archaeological Science 27(1): 1-3.
https://doi.org/10.1006/jasc.1999.0520

Richards, M.P., P.B. Pettitt, M.C. Stiner, and E. Trinkaus. (2001). Stable isotope evidence for increasing dietary breadth in the European Mid-Upper Paleolithic. Proceedings of the National Academy of Sciences of the United States of America 98(11): 6528-6532.
https://doi.org/10.1073/pnas.111155298

Riches, D. (1974). The Netsilik Eskimo: A special case of selective female infanticide. Ethnology 13(4): 351-361.
https://doi.org/10.2307/3773051

Roach, M. (2005). The Roots of Desire. The Myth, Meaning, and Sexual Power of Red Hair. New York: Bloomsbury.

Robins, A.H. (1991). Biological perspectives on human pigmentation. Cambridge: Cambridge University Press.

Rodd, F.H., K.A. Hughes, G.F. Grether, and C.T. Baril. (2002). A possible non-sexual origin of mate preference: Are male guppies mimicking fruit? Proceedings of the Royal Society B 269(1490): 475-81.
https://dx.doi.org/10.1098%2Frspb.2001.1891

Rogers, A.R., R.J. Bohlender, and C.D. Huff. (2017). Early history of Neanderthals and Denisovans. Proceedings of the National Academy of Sciences 114 (37): 9859-9863.
https://doi.org/10.1073/pnas.1706426114

Rogers, R. A. (1986). Language, human subspeciation, and ice age barriers in Northern Siberia. Canadian Journal of Anthropology 5(1): 11-22.

Rosenberg, A., and J. Kagan. (1987). Iris pigmentation and behavioral inhibition. Developmental Psychobiology 20(4): 377-392.
https://doi.org/10.1002/dev.420200403

Rubin, K.H., and L. Both. (1989). Iris pigmentation and sociability in childhood: A re-examination, Developmental Psychobiology 22(7): 717-725
https://doi.org/10.1002/dev.420220706

Rupp, H.A., T.W. James, E.D. Ketterson, D.R. Sengelaub, E. Janssen, and J.R. Heiman. (2009). Neural activation in women in response to masculinized male faces: mediation by hormones and psychosexual factors. Evolution and Human Behavior 30(1): 1-10.
https://doi.org/10.1016/j.evolhumbehav.2008.08.006

Russell, R. (2003). Sex, beauty, and the relative luminance of facial features. Perception 32(9): 1093-1107.
http://dx.doi.org/10.1068/p5101

Russell, R. (2009). A sex difference in facial pigmentation and its exaggeration by cosmetics. Perception 38(8): 1211-1219.
https://doi.org/10.1068/p6331

Russell, R. (2010). Why cosmetics work. In R.B. Adams Jr., N. Ambady, K. Nakayama, and S. Shimojo (Eds.) The Science of Social Vision, (pp. 186-203). New York: Oxford.

Russell, R., B. Duchaine, and K. Nakayama. (2009). Super-recognizers: People with extraordinary face recognition ability. Psychonomic Bulletin & Review 16(2): 252-257.
https://dx.doi.org/10.3758%2FPBR.16.2.252

Russell, R. and P. Sinha. (2007). Real-world face recognition: The importance of surface reflectance properties. Perception 36(9): 1368-1374.
http://dx.doi.org/10.1068/p5779

Russell, R., P. Sinha, I. Biederman, and M. Nederhouser. (2006). Is pigmentation important for face recognition? Evidence from contrast negation. Perception 35: 749-759.
https://doi.org/10.1068%2Fp5490

Saladin d'Anglure, B. (1988) Penser le "féminin" chamanique, Recherches amérindiennes au Québec 18(2-3): 19-50.

Schapera, I. (1966). Married Life in an African Tribe. Evanston: Northwestern University Press.

Scheffel, D. (1984). From polygyny to cousin marriage? Acculturation and marriage in 19th century Labrador Inuit society. Inuit Studies 8(2): 61-75.
https://www.jstor.org/stable/42869417

Schild, R. (1976). The final Paleolithic settlements of the European Plain. Scientific American 234(2): 88-99.
https://doi.org/10.1038/scientificamerican0276-88

Schrire, C., and W.L. Steiger. (1974). A matter of life and death: An investigation into the practice of female infanticide in the Arctic. Man 9(2): 161-184.
https://doi.org/10.2307/2800072

Schweder, B.I.M. (1994). The impact of the face on long-term human relationships. Homo 45(1): 74-93.

Scozzari, R., F. Cruciani, P. Malaspina, P. Santolamazza, B.M. Ciminelli, A. Torroni, et al. (1997). Differential structuring of human populations for homologous X and Y microsatellite loci. American Journal of Human Genetics 61(3): 719-733.
https://doi.org/10.1086/515500

Segrave, K. (2005). Suntanning in 20th Century America. Jefferson (North Carolina): McFarland & Company.

Seki, K. (1963). Folktales of Japan. Chicago: The University of Chicago Press.

Semin, G.R., T. Palma, C. Acartürk, and A. Dziuba. (2018). Gender is not simply a matter of black and white, or is it? Philosophical Transactions of the Royal Society B Biological Sciences 373(1752):20170126
https://doi.org/10.1098/rstb.2017.0126

Shekar, S.N., D.L. Duffy, T. Frudakis, G.W. Montgomery, M.R. James, R.A. Sturm, and N.G. Martin. (2008). Spectrophotometric methods for quantifying pigmentation in human hair-Influence of MC1R genotype and environment. Photochemistry and Photobiology 84(3): 719-726.
https://doi.org/10.1111/j.1751-1097.2007.00237.x

Short, G.B. (1975). Iris pigmentation and phototopic visual acuity: A preliminary study. American Journal of Physical Anthropology 43(3): 425-434.
https://doi.org/10.1002/ajpa.1330430324

Sieber, R. and F. Herreman. (2000). Hair in African Art and Culture. African Arts 33(3): 54-96.
https://doi.org/10.2307/3337689

Siepe, F. (2004). Farben des Eros. Marginalien zur Kulturgeschichte der Liebes- und Schönheitswahrnehmung in Antike und christlichem Abendland. Marburg: Kline.

Sigler, R. (2011). Our Long Hairitage. Bloomington (Indiana): WestBow Press.

Siiteri, P.K. and P.C. MacDonald. (1973). Role of extraglandular estrogen in human endocrinology. In S.R. Geiger (Ed.), Handbook of Physiology, vol. II, Part 1, (pp. 615-629). Washington D.C.: American Physiology Society, Section 7.

Simchuk, A.P. (2001). Frequency-dependent sexual selection in a natural population of oak leafroller moth (Tortrix viridana L.). Tsitologiya i Genetika 35(5): 25-29.

Simcoe, M., A. Valdes, F. Liu, N.A. Furlotte, D.M. Evans, G. Hemani, et al. (2021). Genome-wide association study in almost 195,000 individuals identifies 50 previously unidentified genetic loci for eye color. Science Advances 7(11): eabd1239
https://doi.org/10.1126/sciadv.abd1239

Sinnock, P. (1970). Frequency dependence and mating behavior in Tribolium castaneum. The American Naturalist 104(939): 469-476.
https://doi.org/10.1086/282681

Sitek, A., S. Koziel, A. Kasielska-Trojan, and B. Antoszewski. (2018). Do skin and hair pigmentation in prepubertal and early pubertal stages correlate with 2D:4D? American Journal of Human Biology 30(6): e12631.
https://doi.org/10.1002/ajhb.23183

Skirda, A. (2010). La traite des Slaves. L'esclavage des Blancs du VIIIe au XVIIIe siècle. Paris: Les Éditions de Paris Max Chaleil.

Skoglund, P., H. Malmström, M. Raghavan, J. Storå, P. Hall, E. Willerslev, M.T. Gilbert, A. Götherström, and M. Jakobsson. (2012). Origins and genetic legacy of Neolithic farmers and hunter-gatherers in Europe. Science 336: 466-469.
https://doi.org/10.1126/science.1216304)

Smith, E.A, and S.A. Smith. (1994). Inuit sex-ratio variation. Population control, ethnographic error, or parental manipulation. Current Anthropology 35(5): 595-624.
https://doi.org/10.1086/204319

Smith, M.F. (1965). Baba of Karo: A Woman of the Muslim Hausa. London: Faber & Faber.

Snell, R.S., and R. Turner. (1966). Skin pigmentation in relation to the menstrual cycle. The Journal of Investigative Dermatology 47(2): 147-155.
https://doi.org/10.1038/jid.1966.119

Soejima, M., H. Tachida, T. Ishida, A. Sano, and Y. Koda. (2005). Evidence for recent positive selection at the human AIM1 locus in a European population. Molecular Biology & Evolution 23(1): 179-188.
https://doi.org/10.1093/molbev/msj018

Somigliana, E., P. Viganò, A. Abbiati, D. Gentilini, F. Parazzini, L. Benaglia, et al. (2010). 'Here comes the sun': pigmentary traits and sun habits in women with endometriosis. Human Reproduction 25(3):728-773.
https://doi.org/10.1093/humrep/dep453

Soustelle, J. (1970). The Daily Life of the Aztecs. Stanford, California: Stanford University Press.

Stansfield, J., and T.W.A. Whitfield. (2005) Can future colour trends be predicted on the basis of past colour trends? An empirical investigation. Color Research & Application 30(3): 235-242.
https://doi.org/10.1002/col.20110

Steggerda, M. (1941). Change in hair color with age. Journal of Heredity 32(11): 402-403.
https://doi.org/10.1093/oxfordjournals.jhered.a104977

Stephen, I.D., F.H. Oldham, D.I. Perrett, and R.A. Barton. (2012). Redness enhances perceived aggression, dominance and attractiveness in men's faces. Evolutionary Psychology 10(3)
https://doi.org/10.1177%2F147470491201000312

Sturm, R.A., D.L. Duffy, Z.Z. Zhao, F.P.N. Leite, M.S. Stark, N.K. Hayward, et al. (2008). A single SNP in an evolutionary conserved region within intron 86 of the HERC2 gene determines human blue-brown eye color. American Journal of Human Genetics 82(2): 424-431.
https://doi.org/10.1016/j.ajhg.2007.11.005

Sturm, R.A., and T.N. Frudakis. (2004). Eye colour: portals into pigmentation genes and ancestry. Trends in Genetics 20(8): 327-332.
https://doi.org/10.1016/j.tig.2004.06.010

Suarez, P., K. Baumier, and D. Hall. (2021). Further insight into the global variability of the OCA2-HERC2 locus for human pigmentation from multiallelic markers. Scientific Reports 11(1):22530
https://www.nature.com/articles/s41598-021-01940-w

Sykorová, K., V. Fiala, J. Hlaváčová, S. Kanková, and J. Flegr. (2021). Redheaded women are more sexually active than other women, but it is probably

due to their suitors. OSF preprints, October 13
https://doi.org/10.31219/osf.io/8frq7

Symons, D. (1995). Beauty is in the adaptations of the beholder: The evolutionary psychology of human female sexual attractiveness. In P.R. Abramson and S.D. Pinkerton (Eds.) Sexual Nature. Sexual Culture, (pp. 80-118). Chicago: University of Chicago Press.

Tan, J., Y. Yang, K. Tang, P.C. Sabeti, L. Jin, and S. Wang. (2013). The adaptive variant EDARV370A is associated with straight hair in East Asians. Human Genetics 132(10): 1187-1191.
http://dx.doi.org/10.1007/s00439-013-1324-1

Tarr, M.J., D. Kersten, Y. Cheng, and B. Rossion. (2001). It's Pat! Sexing faces using only red and green. Journal of Vision 1(3): 337, 337a.
https://doi.org/10.1167/1.3.337

Tarr, M. J., B. Rossion, and K. Doerschner. (2002). Men are from Mars, women are from Venus: Behavioral and neural correlates of face sexing using color. Journal of Vision 2(7): 598, 598a,
http://journalofvision.org/2/7/598/, https://doi.org/10.1167/2.7.598.

Tegner, E. (1992). Sex differences in skin pigmentation illustrated in art. The American Journal of Dermatopathology 14(3): 283-287.
https://doi.org/10.1097/00000372-199206000-00016

Tell-Marti, G., J.A. Puig-Butille, M. Potrony, C. Badenas, M. Milà, J. Malvehy, et al. (2015). The MC1R melanoma risk variant p.R160W is associated with Parkinson disease. Annals of Neurology 77(5):889-894.
https://doi.org/10.1002/ana.24373

Tembo, M.S. (2010). The Rediscovery of the Beautiful Woman in African Societies. Eurocentric Destruction of Indigenous Conceptions: the Secret Rediscovery of the Beautiful Woman in African Societies.
https:// wp.bridgewater.edu/mtembo/articles/beautiful-women-in-african-societies/

Templeton, A. (2002). Out of Africa again and again. Nature 416(6876): 45-51.
https://doi.org/10.1038/416045a

The ENCODE Project Consortium. (2012). An integrated encyclopedia of DNA elements in the human genome. Nature 489: 57-74
https://doi.org/10.1038/nature11247

Thelen, T.H. (1983). Minority type human mate preference. Social Biology 30(2): 162-180.
https://doi.org/10.1080/19485565.1983.9988531

Thomas, L.M. (2006). The modern girl and racial respectability in 1930s South Africa. Journal of African History 47(3): 461-490
https://doi.org/10.1017/S0021853706002131

Thompson, S. (1975). Motif-Index of Folk-Literature. Bloomington & London: Indiana University Press.

Thorstenson, C.A. (2018). The social psychophysics of human face color: Review and recommendations. Social Cognition 36(2): 247-273
https://doi.org/10.1521/soco.2018.36.2.247

Thorstenson, C.A., A.D. Pazda, S.G. Young, and A.J. Elliot. (2019a). Face color facilitates the disambiguation of confusing emotion expressions: Toward a social functional account of face color in emotion communication. Emotion, 19(5): 799-807.
https://doi.org/10.1037/emo0000485

Thorstenson, C.A., A. Pazda, and S. Lichtenfeld. (2019b). Facial blushing influences perceived embarrassment and related social functional evaluations. Cognition and Emotion June 23, 1-14.
https://doi.org/10.1080/02699931.2019.1634004

Thorstenson, C.A., J. McPhetres, A.D. Pazda, and S.G. Young. (2021). The role of facial coloration in emotion disambiguation. Emotion. Advance online publication.
https://doi.org/10.1037/emo0000900

Tong, K., K.A. Hughes, and A.E. Houde. (2009). The allure of the distinctive: reduced sexual responsiveness of female guppies to 'redundant' male colour patterns. Ethology 115(5): 475-481.
https://doi.org/10.1111/j.1439-0310.2009.01634.x

Torroni, A., O. Semino, R. Scozzari, G. Sirugo, G. Spedini, N. Abbas, et al. (1990). Y-chromosome DNA polymorphisms in human populations: Differences between Caucasoids and Africans detected by 49a and 49f probes. Annals of Human Genetics 54(4): 287-296.
https://doi.org/10.1111/j.1469-1809.1990.tb00384.x

Trivers, R., J. Manning, and A. Jacobson. (2006). A longitudinal study of digit ratio (2D:4D) and other finger ratios in Jamaican children. Hormones and Behavior 49(2): 150-156. https://doi.org/10.1016/j.yhbeh.2005.05.023

Unterländer, M., F. Palstra, I. Lazaridis, A. Pilipenko, Z. Hofmanová, M. Gross, et al. (2017). Ancestry and demography and descendants of Iron Age nomads of the Eurasian Steppe. Nature Communications 8(14615) http://dx.doi.org/10.1038/ncomms14615

Valković, V. (1988). Human Hair. Vol. 1. Fundamentals and Methods for Measurement of Elemental Composition. Boca Raton (Florida): CRC Press.

Vallois, H.V. (1962). The social life of early man: The evidence of skeletons. In S.I Washburn (Ed.) Social Life of Early Man (pp. 214-235). Viking Fund Publications in Anthropology 31. https://doi.org/10.2753/aae1061-1959010454

van den Berghe, P.L. (1979). Human family systems. An evolutionary view. New York: Elsevier.

van den Berghe, P.L., and P. Frost. (1986). Skin color preference, sexual dimorphism and sexual selection: A case of gene-culture co-evolution? Ethnic and Racial Studies 9(1): 87-113. https://doi.org/10.1080/01419870.1986.9993516

Vasvari, L.O. (1999). A comparative approach to European folk poetry and the erotic wedding motif. CLCWeb. Comparative Literature and Culture 1(4) https://doi.org/10.7771/1481-4374.1050

Vercoutter, J., J. Leclant, F.M. Snowden Jr., and J. Desanges. (1976). The Image of the Black in Western Art, vol. 1. Menil.

Verlinden, C. (1977). L'Esclavage dans l'Europe médiévale, vol. II. Ghent.

Vilakazi, A. (1962). Zulu Transformations. Pietermaritzburg: University of Natal Press.

Visscher, M.O., S.A. Burkes, D.M. Adams, A.M. Hammill, and R.R. Wickett. (2017). Infant skin maturation: preliminary outcomes for color and biochemical properties. Skin Research & Technology 23(4): 545-551. https://doi.org/10.1111/srt.12369

Voight, B.F., S. Kudaravalli, X. Wen, and J.K. Pritchard. (2006). A map of recent positive selection in the human genome. PLoS Biology 4(3): e72. https://doi.org/10.1371/journal.pbio.0040072

Wagatsuma, H. (1967). The social perception of skin color in Japan. Daedalus 96(2): 407-443. http://www.jstor.org/stable/20027045

Walentowitz, S. (2008). Des êtres à peaufiner. Variations de la coloration et de la pigmentation du nouveau-né. In J-P. Albert, B. Andrieu, P. Blanchard, G. Boëtsch, and D. Chevé (Eds.) Coloris Corpus, (pp. 113-120). Paris: CNRS Éditions.

Walsh, R. J. (1964). Variation in the melanin content of the skin of New Guinea natives at different ages. Journal of Investigative Dermatology 42(3): 261-265. https://doi.org/10.1038/jid.1964.59

Walsh, R.J., and R.E. Chapman. (1966). A study of the quantitative measurement of human head hair fibres. Man, new series. 1(2): 226-232. https://doi.org/10.2307/2796348

Walsh, R.J., F. Liu, A. Wollstein, L. Kovatsi, A. Ralf, A. Kosiniak-Kamysz., et al. (2013). The HIrisPlex system for simultaneous prediction of hair and eye colour from DNA. Forensic Science International: Genetics 7: 98-115. http://dx.doi.org/10.1016/j.fsigen.2012.07.005

Weinbaum, A.E., L.M. Thomas, P. Ramamurthy, U.G. Poiger, M.Y. Dong, and T.E. Barlow (eds.) (2008). The Modern Girl Around the World. Consumption, Modernity, and Globalization. Durham and London: Duke University Press. https://doi.org/10.1215/9780822389194

Weinhold, K. (1882). Die Deutschen Frauen in dem Mittelalter. Vienna: Druck und Verlag von Carl Gerald's Sohn.

Wellenreuther, M., E.I. Svensson, and B. Hansson. (2014). Sexual selection and genetic colour polymorphisms in animals. Molecular Ecology 23(22): 5398-5414. https://doi.org/10.1111/mec.12935

West, M.L. (1996) Hesiod. Works & Days. Oxford: University Press.

Weyer, E.M. (1932). The Eskimos. New Haven, Conn.: Yale University Press.

White, C.M.M. (1954). Correspondence. Man 54: 147.

White, S. and G. White. (1995). Slave Hair and African American Culture in the Eighteenth and Nineteenth Centuries. The Journal of Southern History 61(1): 45-76. https://doi.org/10.2307/2211360

White, T.E. (2017). Digest: Strengthening the link between sexual selection and color polymorphism. Evolution 71(7): 1913-1914. https://doi.org/10.1111/evo.13272

Wickler, W. (1973). The Sexual Code. Garden City: Anchor.

Williams, J.E., D.A. Boswell, and D.L. Best. (1975). Evaluative responses of preschool children to the colors white and black. Child Development 46(2): 501-508.
https://doi.org/10.2307/1128148

Williams, J.E., D.L. Best, D.A. Boswell, L.A. Mattson, and D.J. Graves. (1975). Preschool racial attitude measure II. Educational and Psychological Measurement 35(1): 3-18.
https://doi.org/10.1177/001316447503500101

Williams, J.E. and J.K. Roberson. (1967). A method for assessing racial attitudes in preschool children. Educational and Psychological Measurement, 27(3): 671-689.
https://doi.org/10.1177/001316446702700310

Williams, J.E. and C.A. Rousseau. (1971). Evaluation and identification responses of Negro preschoolers to the colors black and white. Perceptual and Motor Skills 33(2): 587-599.
https://doi.org/10.2466/pms.1971.33.2.587

Wimberly, L.C. (1929). The red-headed man. Prairie Schooner 3(4): 284-290.
http://www.jstor.org/stable/40622087

Woodworth, S.H., J.S. Sanfilippo, M. Singh, C.L. Cook, M.A. Yussman, and S.R. Lincoln. (1995). A prospective study on the association between red hair color and endometriosis in infertile patients. Fertility and Sterility 64(3):651-652.
https://doi.org/10.1016/S0015-0282(16)57809-1

Woolf, C.M. (1965). Albinism among Indians in Arizona and New Mexico. American Journal of Human Genetics 17(1): 23-35.

Woolf, C.M. and F.C. Dukepoo. (1969). Hopi Indians, inbreeding, and albinism. Science 164(3875): 30-37.
https://doi.org/10.1126/science.164.3875.30

Yip, A.W., and P. Sinha. (2002). Contribution of color to face recognition. Perception 31(8): 995-1003.
https://doi.org/10.1068/p3376

Zahan, D. (1974). White, Red and Black: Colour Symbolism in Black Africa. In A. Portmann and R. Ritsema (Eds.) The Realms of Colour, Eranos 41 (1972): 365-395, Leiden: Eranos.
https://doi.org/10.2307/1573594

Zhu, Q., Y. Song, S. Hu, X. Li, M. Tian, Z. Zhen, Q. Dong, N. Kanwisher, and J. Liu. (2009). Heritability of the specific cognitive ability of face perception. Current Biology 20(2): 137-142.
https://doi.org/10.1016/j.cub.2009.11.067

Zuckerman, M. (2008). Genetics of Sensation Seeking. In J. Benjamin, R.P. Ebstein, and R.H. Belmaker (Eds) Molecular Genetics and the Human Personality, (pp. 193-210). Washington D.C.: American Psychiatric Publishing Inc.

Index

www.ingramcontent.com/pod-product-compliance
Lightning Source LLC
LaVergne TN
LVHW020414220926
848484LV00018B/2580/J
* 9 7 8 1 6 8 0 5 3 8 7 2 4 *